Smart Innovation, Systems and Technologies

Volume 304

Series Editors

Robert J. Howlett, Bournemouth University and KES International, Shoreham-by-Sea, UK

Lakhmi C. Jain, KES International, Shoreham-by-Sea, UK

The Smart Innovation, Systems and Technologies book series encompasses the topics of knowledge, intelligence, innovation and sustainability. The aim of the series is to make available a platform for the publication of books on all aspects of single and multi-disciplinary research on these themes in order to make the latest results available in a readily-accessible form. Volumes on interdisciplinary research combining two or more of these areas is particularly sought.

The series covers systems and paradigms that employ knowledge and intelligence in a broad sense. Its scope is systems having embedded knowledge and intelligence, which may be applied to the solution of world problems in industry, the environment and the community. It also focusses on the knowledge-transfer methodologies and innovation strategies employed to make this happen effectively. The combination of intelligent systems tools and a broad range of applications introduces a need for a synergy of disciplines from science, technology, business and the humanities. The series will include conference proceedings, edited collections, monographs, handbooks, reference books, and other relevant types of book in areas of science and technology where smart systems and technologies can offer innovative solutions.

High quality content is an essential feature for all book proposals accepted for the series. It is expected that editors of all accepted volumes will ensure that contributions are subjected to an appropriate level of reviewing process and adhere to KES quality principles.

Indexed by SCOPUS, EI Compendex, INSPEC, WTI Frankfurt eG, zbMATH, Japanese Science and Technology Agency (JST), SCImago, DBLP.

All books published in the series are submitted for consideration in Web of Science.

More information about this series at https://link.springer.com/bookseries/8767

Yiming Bie · Bob X. Qu ·
Robert J. Howlett · Lakhmi C. Jain
Editors

Smart Transportation Systems 2022

Proceedings of 5th KES-STS International Symposium

Springer

Editors
Yiming Bie
School of Transportation
Jilin University
Changchun, China

Bob X. Qu
Chalmers University of Technology
Göteborg, Sweden

Robert J. Howlett
KES International
Shoreham-by-Sea, UK

Lakhmi C. Jain
KES International
Shoreham-by-Sea, UK

ISSN 2190-3018 ISSN 2190-3026 (electronic)
Smart Innovation, Systems and Technologies
ISBN 978-981-19-2812-3 ISBN 978-981-19-2813-0 (eBook)
https://doi.org/10.1007/978-981-19-2813-0

This Springer imprint is published by the registered company Springer Nature Singapore Pte Ltd.
The registered company address is: 152 Beach Road, #21-01/04 Gateway East, Singapore 189721, Singapore

Preface

Transport systems are evolving through advanced technologies in automation, communication, energy, and computation power. These new technologies provide promising and great potential to improve efficiency (both travel and energy), accessibility, resilience, and connectivity of current mobility services, and thus facilitate a healthy transition to a sustainable transport system and climate-neutral mobility. To name a few, electric vehicles are sustainable alternatives to reduce fossil energy consumption and greenhouse emission in the transport sector and have been strongly promoted worldwide in the past years; connected and automated vehicles have made notable progress in the past decade and are being tested and demonstrated by various stakeholders such as car manufacturers and travel service providers; big transport data thanks to emerging information, and connected technologies enable more comprehensive analysis about key components in transport systems and bring a prosperity of machine learning methods to conduce powerful analysis in a data-driven way. However, the benefits of new technologies are always accompanied by challenges, which require collective efforts among academic researchers and practitioners. Breakthroughs in methodologies and applications are required to fully leverage the abovementioned techniques for a more sustainable future transport system. In this regard, the Fifth International Symposium on Smart Transport Systems is going to be organized in June 2022 to provide a favorable communication environment, exchanging knowledge and collaborative platform among researchers and practitioners in the fields of smart transport systems. The vision is to facilitate complementary collaborations among academic and industrial communities in terms of research, implementation, and applications.

In the symposium of 2022, 19 excellent papers from several countries were finally accepted for the proceeding. These papers were rigorously peer-reviewed in two rounds by at least two independent external reviewers and one editorial member. A kind and joint dialog between reviewers and authors have been established for improving these papers from different aspects. We have received studies about several interesting topics in the scope of smart transport systems. We have received two interesting and sound studies about electric bus scheduling optimization and operation management to improve efficiency and reduce overall cost. Another two studies address the human factors and driving behavior reflections regarding advanced driving-assistance systems and connected vehicles with a focus on car-following behavior. Two studies investigate the sharing mobility, including demand analysis and prediction of car sharing, and usage pattern analysis of e-scooter sharing systems based on big data. Another five studies focus on traffic safety analysis and behavior modeling in terms of human drivers' speeding behavior, hard-braking behavior, and traffic violations, as well as scenario-oriented design for the safety of autonomous vehicles. Field data are used for analysis and to derive countermeasures for improvement strategy. Moreover, we have collected three interesting studies about traffic flow theory, modeling and control in different scenarios including highway, roundabouts and connected intersections. Another three articles offer exciting and applicable methods of utilizing multisource data and data-driven algorithms for anomaly detection, road pavement material recognition, and dynamic imputation. The rest two studies shed light on route choice behavior based on experimental data and emission modeling of different vehicles in intersections, respectively. Participants and authors are mainly scholars and practitioners from Sweden, Germany, Belgium, and Australia, who have delivered their interesting research outcomes, exchanged opinions, and built networks during the symposium with a pleasant and relaxing atmosphere.

Contents

About the Editors

Yiming Bie is Professor of Transportation Engineering with the School of Transportation, Jilin University, China. He obtained his Ph.D. degree at Jilin University in June 2012. His research interests include public transportation operations, traffic control, and intersection design. He has authored or co-authored over 70 journal articles at top-tier journals such as Transportation Research Part C, Computer-Aided Civil and Infrastructure Engineering, Journal of Transportation Engineering-ASCE. The algorithms developed by him were adopted by the most popular adaptive traffic control system in China and has been implemented to more than 20 cities. Recently, his research is focused on electric bus operations in cold regions. He is Principal Investigator for two projects funded by National Natural Science Foundation of China and a few projects from other funding agencies. In 2019, he was conferred the outstanding reviewer recognition by Journal of Transportation Engineering-ASCE.

Bob X. Qu is Professor and Research Group Leader at the Division of Geology and Geotechnics, Chalmers University of Technology. Throughout his academic career, he has been endeavoring to practically improve transport safety, efficiency, equity, and sustainability through traffic flow modeling, network optimization, and most recently emerging technologies. In particular, his research has been applied to improvement of emergency services, operations of electric vehicles and connected automated vehicles, and management of vulnerable road users. He has authored or co-authored over 90 journal articles published at top-tier journals in the area of transport engineering, and he is a recipient of many prestigious awards. His research has been supported by Australian Research Council Discovery Programme, Queensland Department of Transport and Main Roads, Sydney Trains, National Natural Science Foundation of China, Swedish Innovation Agency Vinnova, and European Union.

Dr. Robert J. Howlett is Executive Chair of KES International, a non-profit organization that facilitates knowledge transfer and the dissemination of research results in areas including Intelligent Systems, Sustainability, and Knowledge

Transfer. He is Visiting Professor at Bournemouth University in the UK. His technical expertise is in the use of intelligent systems to solve industrial problems. He has been successful in applying artificial intelligence, machine learning, and related technologies to sustainability and renewable energy systems; condition monitoring, diagnostic tools and systems; and automotive electronics and engine management systems. His current research work is focused on the use of smart microgrids to achieve reduced energy costs and lower carbon emissions in areas such as housing and protected horticulture.

Dr. Lakhmi C. Jain, Ph.D., ME, BE(Hons), Fellow (Engineers Australia), is with the University of Technology Sydney, Australia, and Liverpool Hope University, UK. She serves the KES International for providing a professional community the opportunities for publications, knowledge exchange, cooperation, and teaming. Involving around 5,000 researchers drawn from universities and companies worldwide, KES facilitates international cooperation and generates synergy in teaching and research. KES regularly provides networking opportunities for professional community through one of the largest conferences of its kind in the area of KES.

Complexity Quantification of Car-Following Dynamic Traffic in the Internet of Vehicles Environment

Yaoyin Zhang[1], Linhong Wang[1], and Ce Wang[2](✉)

[1] Transportation College of Jilin University, Changchun 130022, Jilin, China
[2] Changchun FAWSN Automotive Technology Research and Development Co., Ltd., Changchun 130103, Jilin, China
cwang07@fawsn-autoparts.com

Abstract. The complexity of the car-following dynamic traffic means the driver's workload brought by the main driving task during the car-following process. The status information between the preceding and following vehicles on the Internet of Vehicles environment provides real-time quantification of the complex dynamic traffic environment. In order to quantify the complexity of the traffic environment in real time, three dynamic traffic environment complexity metrics based on the car-following behavior spectrum are used, which are Inversed Modified Margin to Collision, Transverse Oscillation Coefficient and Velocity Instability Coefficient. And we use the driving simulator to collect driver's behavior data under different driving tasks during car-following, calculate the specific values of the three metrics, and use the entropy method to take a weighted sum of the three metrics. At last, we also use NASA-TLS subjective load scale to obtain the driver's subject load, and this can verify the quantification model of the car-following dynamic traffic environment. The results show that the complexity quantification model of the car-following traffic environment can accurately quantify the driver's workload brought by the main driving task to the driver during the car-following process. The study will provide a scientific basis for the control switching of vehicles between driver and machine and supply a more efficient driving behavior.

Keywords: Traffic environment complexity · Internet of vehicle · Car-following behavior spectrum · Driver's workload

1 Introduction

With the increase of car ownership, the traffic environment has become more and more complex. Drivers always pay attention to the movement state of the surrounding vehicles and adjust their driving behaviors when they drive the vehicle. The increasingly complex dynamic traffic environment requires drivers more energy to focus on driving tasks. However, due to people's limited capabilities of dealing with information and collecting resources, drivers' information processing capabilities are not sufficient to deal with the extremely complex dynamic traffic. The complex environment leads to

Y. Bie et al. (Eds.): KES-STS 2022, SIST 304, pp. 1–10, 2022.
https://doi.org/10.1007/978-981-19-2813-0_1

the driver's driving workload increase, and it is easy to affect the driving performance. Previous studies have shown that the complex dynamic traffic environment will increase the driver's perception of driving difficulty, which will increase the driver's workload, affect driving performance, and thus have an impact on road traffic safety [1]. With the development of the Internet of Vehicles technology, vehicles can obtain information about other vehicles on the road [2], and then can quantify the complex vehicle following process in real time, which is of great significance for reducing the driving workload of drivers and improving driving safety.

To measure traffic environment complexity, many efforts have already been made previously. For example, some researchers used the visual information of the driver to quantify the complexity of the road traffic environment, and used image processing methods to propose a calculation method of the complexity of the traffic environment based on the calculation of the driver's visual information [3]. By calculating the driver's visual attention to the road information, the road traffic environment was divided into different layers, and the weight coefficient of each visual environment layer was determined by the eye movement data, and an information calculation model based on the visual information hierarchical model was established [4]. Another part of the scholars used driving simulators to simulate different driving scenarios. Through the analysis and processing of the driver's blinking frequency, electrocardiogram and other physiological responses and driving performance, they explored the impact of different complex levels of traffic scenarios on the driver's state and driving behavior to evaluate the complexity of the road traffic environment [5–12]. Some scholars also proposed metrics to quantify the complexity of the dynamic car-following traffic environment. Xue et al. used metrics such as speed, acceleration, reciprocal collision time, headway time, and other metrics, and used the modified margin to the collision to evaluate the risk of car-following behavior for different vehicle types [13]. He and others used speed, acceleration, accelerator/brake pedal force, head distance, relative speed, TTC, and other metrics as the components of the car-following behavior spectrum. They constructed the car-following behavior spectrum of 6 types of drivers in 15 car-following driving scenarios [14].

To quantify the complexity of the dynamic car-following traffic environment in real time, we select three complexity evaluation metrics based on the car-following behavior spectrum, obtain the experimental data of the four car-following tasks performed by the driving simulator, and use the entropy method to weight and sum the three metrics. The complexity quantification model of the car-following dynamic traffic environment is established, and the model is verified by the subjective load of the drivers.

2 Model

2.1 Quantitative Metrics of Car-Following Dynamic Traffic Complexity

During the car following process, there are many situations that will increase the driver's workload, such as emergency braking of the preceding vehicle, serpentine driving of the vehicle, and unstable driving of the vehicle. Based on the three driving situations, the metrics of car-following dynamic traffic complexity are respectively proposed.

IMMC (Inversed Modified Margin to Collision) describes the reciprocal of the following car's driver reaction time before braking, when the preceding car brakes suddenly, which can objectively quantify the driver's workload caused by driving too fast or following the preceding car too close.

$$I_{IMMTC} = \frac{2av_f}{v_p^2 - v_f^2 + 2aD_0} \tag{1}$$

where I_{IMMTC} is the evaluation metric of Inversed Modified Margin to Collision. V_p and V_f are the vehicle speeds of the preceding and following vehicles, respectively. D_0 is the initial distance between the two cars. a is the braking deceleration, which is -8 m/s^{-2}.

Some emergencies (such as road obstacles and road depressions) will cause drivers to adjust the lateral position of the car to avoid it, forming a serpentine driving, and increasing the drivers' driving workload. TOC (Transverse Oscillation Coefficient) is used to evaluate the complexity of snake-like car-following driving.

$$I_{TOC} = \frac{\sum |W(t)|}{d(t) \times F} \tag{2}$$

where I_{TOC} is the evaluation metric of the Transverse Oscillation Coefficient. *d(t)* is the vertical travel distance within 1s before time *t*. *W(t)* is the cumulative value of lateral offset within 1s before time *t*. *F* is the data recording frequency.

Unstable speed driving affects not only driving safety, but also affects driving comfort and consumes drivers' energy. VIC (Velocity Instability Coefficient) is used as an indicator of the complexity of the speed instability.

$$I_{VIC} = \frac{V_{std}(t)}{V_{mean}(t)} \times 100\% \tag{3}$$

where I_{VIC} is the evaluation metric of Velocity Instability Coefficient. $V_{std}(t)$ is the speed standard deviation within 1s before time *t*. $V_{mean}(t)$ is the mean value of speed within 1s before time *t*.

2.2 The Establishment of the Quantitative Model

(1) Metrics weight calculate --- entropy weight method. The entropy method determines the metric weights according to the degree of variation of each metric [15]. This is an objective weighting method that avoids the deviation caused by human factors. Relative to the subjective assignment method, this method has higher accuracy and stronger objectivity and can better explain the results obtained. The weights of I_{IMMTC}, I_{TOC}, and I_{VIC} are set to W_1, W_2, W_3.

(2) Metrics min-max standardization. After statistical calculation: The value ranges of I_{IMMTC}, I_{TOC}, and I_{VIC} are between 0–2 s^{-1}, 0–0.01, 0–3.2, respectively. Therefore, the data needs to be normalized, that is, to transform the original data linearly, and the result value is mapped to [0,1]. The conversion function is as follows:

$$x_{ij}^* = \frac{x_{ij} - min}{max - min} \tag{4}$$

(3) Weighted optimization model. By integrating the various metrics of the complex measurement of the car-following dynamic traffic environment, the complexity quantification model of the car-following dynamic traffic environment is established:

$$C = W_1 \frac{I_{IWWTC}}{2s^{-1}} + W_2 \frac{I_{TOC}}{0.01} + W_3 \frac{I_{VIC}}{3.2} \tag{5}$$

where C is the complexity of the car-following dynamic traffic environment.

3 Experiment Plan and Data Processing

3.1 Experimental Design

Purpose. To quantify the unknown coefficients in the model, this experiment simulates different complexity of car-following scenes on a driving simulator, and the experimenters complete their driving tasks in a virtual traffic environment.

Experiment Plan. Through the UC-winRoad scenario modeling software, a virtual highway road scenario is built in the driving simulator. In this scenario, the road is a two-way six-lane three-level highway, and the lane width is 3.5 m. The road speed limit is 80 km/h. Before the experiment, the running plan of the preceding vehicle is set. Experimenters drive the vehicle to follow the preceding vehicle in the middle lane, and there is traffic flow in other lanes (Fig. 1).

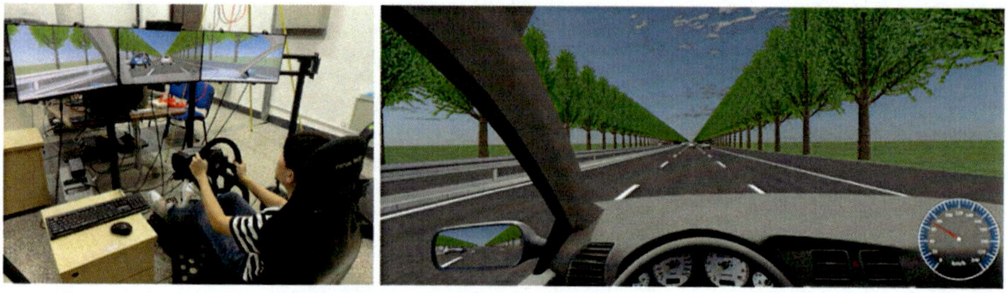

Fig. 1. Simulated driving scenario.

In the experiment, 10 experimenters aged 24–45 years old and more than 5 years of driving experience are selected (Table 1).

Experiment process: In order to record different dynamic traffic environment complexity during the experiment, this study sets four car-following driving tasks, and each task is carried out in sequence. Each experimenter practices using the driving simulator before starting the experiment until they are proficient in operating the driving simulator. The experimenters will fill in the subjective complexity scale after completing a task. The subjective complexity scale is the NASA-TLX evaluation scale developed by the National Aeronautics and Space Administration [16].

Table 1. Basic statistics of drivers.

Statistics	Age	Driving experience/years	Driving mileage/10000 km
Mean	28.6	8.2	8.5
Min	24.0	5.0	0.6
Max	45.0	23.0	14.0

Task 1: The preceding vehicle travels at a constant speed of 40, 50, 60, 70, and 80 km/h on the middle lane, respectively, and each speed lasts for two minutes. Experimenters drive vehicles to follow the preceding vehicle according to their own driving habits.

Task 2: The preceding vehicle travels at a constant speed of 40, 50, 60, 70, and 80 km/h on the middle lane, each speed lasts for two minutes. Experimenters drive vehicles in a serpentine route, and maximize their vehicle's lateral movement without wheels touching the line.

Task 3: The preceding vehicle drives at a constant speed of 40, 50, 60, 70, and 80 km/h on the middle lane, and performs two emergency braking at random. Experimenters drive vehicle to follow the preceding vehicle according to their own driving habits, and they are not allowed to impact the preceding car.

Task 4: The preceding vehicle is driving at a variable speed in the middle lane, and the speed range is random within 30–80 km/h. Experimenters drive car to follow the preceding vehicle according to their own driving habits. In this task, experimenters are required to follow the preceding vehicle as close as possible.

3.2 Data Processing

Find the Weight of Each Indicator. The driving simulator records the driving data of drivers when performing driving tasks, and the data collection frequency of the simulator is 10 Hz. The specific values of 3 metrics will be calculated (Table 2).

Table 2. Fragments of complexity quantification metrics.

I_{IMMTC}	I_{TOC}	I_{VIC}
5.828×10^{-1}	3.017×10^{-6}	1.527×10^{-2}
5.844×10^{-1}	4.399×10^{-6}	1.641×10^{-2}
5.866×10^{-1}	4.366×10^{-6}	1.731×10^{-2}
5.885×10^{-1}	4.354×10^{-6}	1.801×10^{-2}
5.913×10^{-1}	5.822×10^{-6}	1.852×10^{-2}
5.957×10^{-1}	5.726×10^{-6}	1.888×10^{-2}

In order to ensure the universality of the experimental samples, 1000 sets of quantitative metric data are randomly selected from each task of each experimenter, and a total of

32000 sets of quantitative metric data from 8 experimenters are selected (the driving data of the remaining two experimenters were used for verification). W_1, W_2, and W_3 are calculated by entropy weight method and brought into Eq. 5:

$$C = 0.0432 I_{IMMTC} + 26.1000 I_{TOC} + 0.2039 I_{VIC} \tag{6}$$

Determine the Threshold. Considering that most cars nowadays are equipped with electronic body stabilization systems, cars generally do not have risks such as sideslip and overturn when accelerating and decelerating on normal roads and in snake-like car-following. Therefore, I_{IMMTC} is used as the evaluation index to determine the threshold of the complexity of the dynamic car-following traffic environment. The shortest braking reaction time of the driver is 0.5 s [17]. When I_{IMMTC} is greater than 2 s^{-1}, there is a risk of rear-end collision. If I_{TOC} and I_{VIC} are set to 0, $C = 0.0864$. The weight of each metric is enlarged in equal proportion, so that the result of the model is ten at this time:

$$C = 5.00 I_{IMMTC} + 3020.83 I_{TOC} + 23.60 I_{VIC} \tag{7}$$

The complexity of the car-following dynamic traffic environment when the experimenters performed the driving task is quantified by the above model. By analyzing the complexity of the traffic environment under normal driving, the quantile value of complexity 10 can be calculated, which is 95.34% and is judged to be a third-level risk state according to the car-following risk status classification theory, which is only lower than the fourth-level risk status of 99% quantile [18]. Therefore, a state with a complexity of more than 10 is defined as an extremely complex state (Fig. 2).

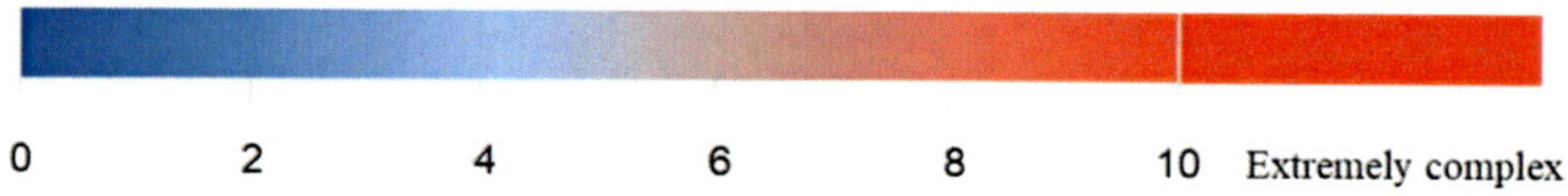

Fig. 2. The complexity level of the dynamic traffic environment.

3.3 Result Analysis

Analysis of Complexity Quantification Results of Dynamic Traffic Environment for Each Task. By using the car-following dynamic traffic environment complexity quantification model, the other two experimenters' results are shown in Fig. 3. Figure 3 shows that the complexity of the dynamic traffic environment of task 1 is around 3, which is in a low-complexity state. When the experimenters perform task 2 with the addition of serpentine driving, the complexity of the dynamic traffic environment is on a high level. In task 3, the preceding vehicle randomly performs emergency braking in order to avoid a collision. The experiment vehicle performs emergency braking in the shortest time. At this time, the complexity of the dynamic traffic environment rises sharply. In task 4, the variable-speed driving of the experimental vehicle causes large fluctuations in the complexity of the dynamic traffic environment, and the overall complexity is at a relatively high level. The mean value of the dynamic traffic environment complexity of each task is shown in Fig. 4. Figure 4 shows that the average complexity of the dynamic traffic environment of driving task 1 is the smallest, followed by driving tasks 2 and 3, and driving task 4 is the largest. The mean value of tasks 1 and 3 gradually grow with the increase of vehicle speed, and the mean value of task 2 first decreases and then increases. This fluctuation should be a small error caused by the driver's driving style. The mean value of the two experiments of task 4 fluctuates slightly, and both are in a high state.

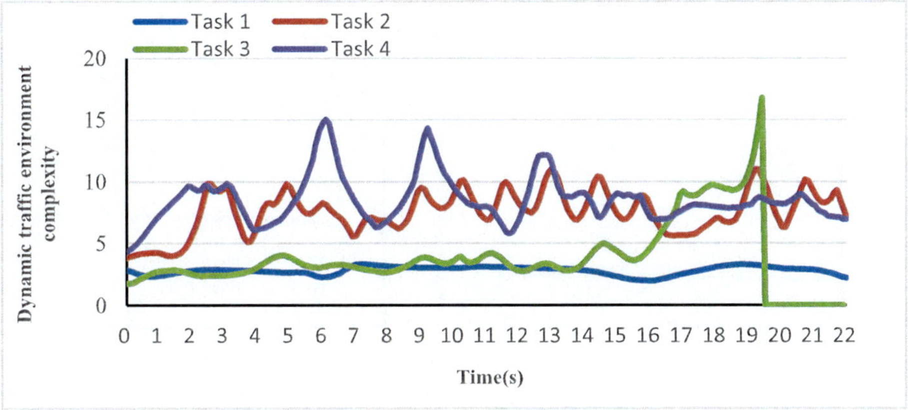

Fig. 3. Fragments of the complexity of the dynamic traffic environment.

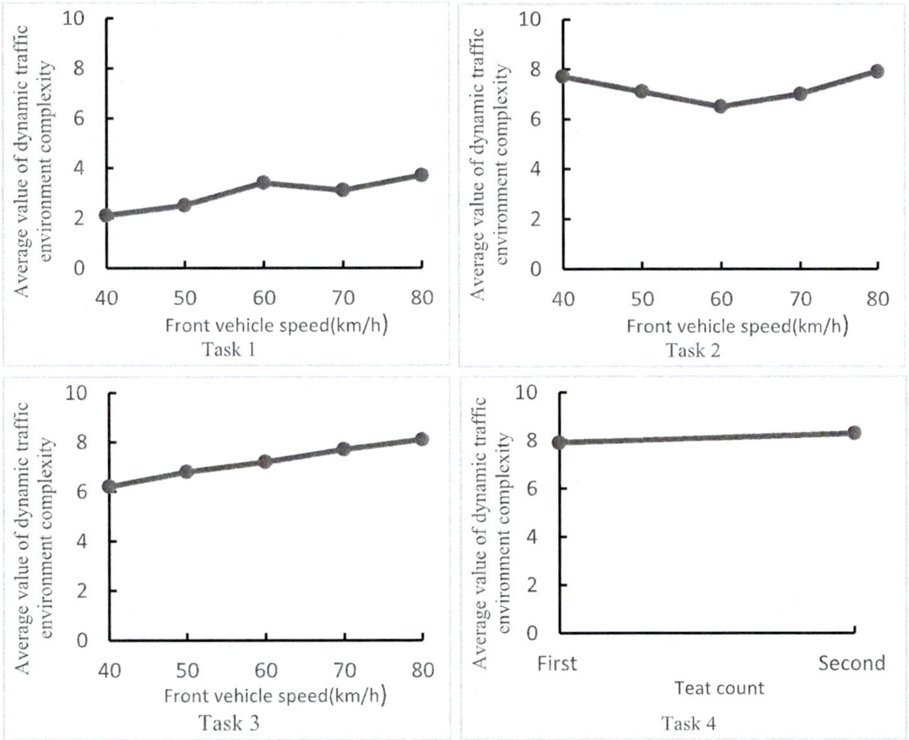

Fig. 4. The average complexity of the dynamic traffic environment for each driving task.

Statistical Analysis of Driver's Workload Subjective Scale. Since the complexity of the traffic environment is positively correlated with the driver's load [6], We compare each task result with the driving workload subjective scale to verify the model. Through data processing of the NASA-TLX subjective load scale, we obtain the subjective load of each driving task of the remaining two experimenters. The subjective load of one of the experimenters is shown in Fig. 5.

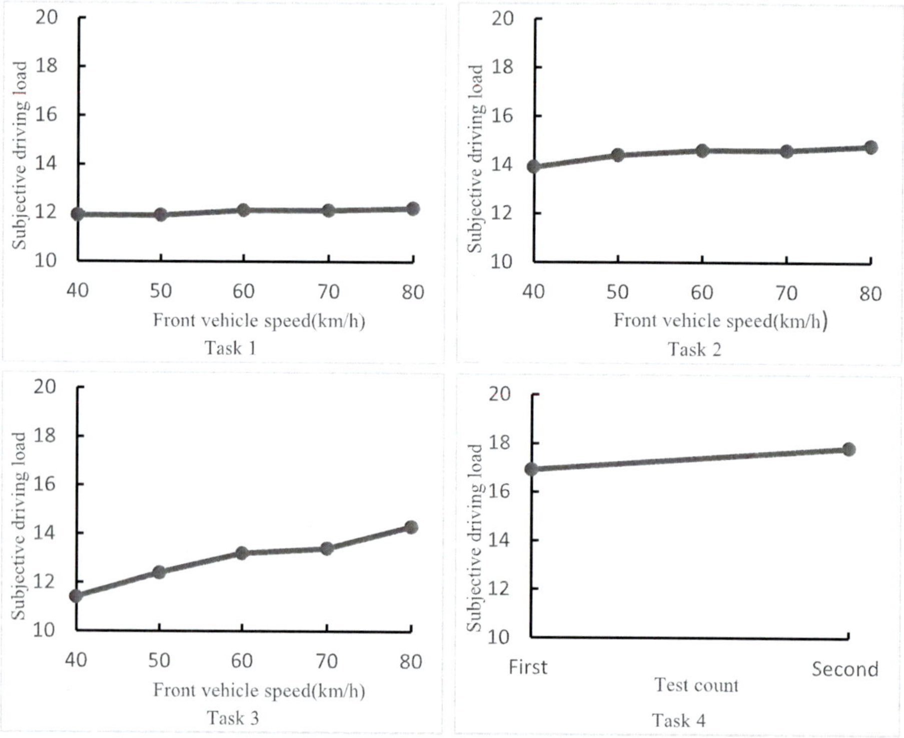

Fig. 5. Subjective driving load of each driving task.

Figure 5 indicates that the subjective load of driving task 1 is the smallest, followed by driving tasks 2 and 3, and task 4 is the largest. The subjective load of driving tasks 2 and 3 slowly grow with the increase of vehicle speed, and the subjective load of driving task 3 is more obvious with the increase of vehicle speed. The subjective load of the two experiments of driving task 4 is at a relatively high level.

Model Validation. Pearson correlation analysis is performed on the average value of the dynamic traffic environment complexity and the subjective load of each driving task of the two experimenters [19], and the two were significantly correlated when the confidence interval is 0.01. The correlation coefficients between the test result and the subjective load of the two experimenters were 0.762 and 0.786, respectively, which shows a strong correlation. It also shows that the car-following dynamic traffic environment complexity quantification model in this thesis can accurately quantify the complexity of the road traffic on which the vehicle is travelling.

4 Conclusion

This thesis establishes a complexity quantification model of the car-following dynamic traffic in the environment of the Internet of Vehicles. Based on the car-following behavior spectrum, three complex metrics are selected: I_{IMMTC}, I_{TOC} and I_{VIC}. Based on the entropy weight method, we took a weighted sum of the 3 metrics and established a complexity quantification model of the car-following dynamic traffic environment. Also, we designed 4 driving tasks in car-following states through the driving simulator and analyzed the driving data. At last, we verified the correlation between the average complexity of the dynamic traffic environment and the subjective load with the Pearson correlation coefficient. The results show that the two are significantly related. Therefore, the driving data obtained in the Internet of Vehicles environment can accurately quantify the complexity of road traffic in the process of the car following in real time. In future research, the complexity quantification of the dynamic traffic environment also needs to be modeled in other driving scenarios to increase the applicating scope. In addition, the authenticity of the research results needs to be verified in real vehicles under safe conditions.

Acknowledgments. This study was supported by the National Key R&D Program of China under Grant (No. 2018YFB1600501), National Natural Science Foundation of China under Grant (No. 71971097), and Youth Program of National Natural Science Foundation of China (No. 52002143).

References

1. Zeitlin, L.R.: Estimates of driver mental workload: a long-term field trial of two subsidiary tasks. Hum. Factors **37**(3), 611–621 (1995)
2. Lim, K.L., Whitehead, J., Jia, D., Zheng, Z.: State of data platforms for connected vehicles and infrastructures. Commun. Transp. Res. **1**, 100013 (2021). https://doi.org/10.1016/j.commtr.2021.100013
3. Wang, S., Pan, X., Zhou, W.: Calculation and analysis of driver's visual information in bridge and tunnel section of expressway. Hunan Commun. Sci. Technol. **37**(4), 156–158 (2011)
4. Dong, Y., Chen, Y.: Evaluation of urban road traffic environment based on the calculation of visual information. J. Transp. Inf. Saf. **32**(6), 146–152 (2014)
5. Faure, V., Lobjois, R., Benguigui, N.: The effects of driving environment complexity and dual tasking on drivers' mental workload and eye blink behavior. Transp. Res. Part F: Traffic Psychol. Behav. **40**, 78–90 (2016)
6. Manawadu, U.E., Kawano, T., Murata, S., Kamezaki, M., Sugano, S.: Estimating driver workload with systematically varying traffic complexity using machine learning: experimental design. In: Karwowski, W., Ahram, T. (eds.) IHSI 2018. AISC, vol. 722, pp. 106–111. Springer, Cham (2018). https://doi.org/10.1007/978-3-319-73888-8_18
7. Sugiono, S., Widhayanuriyawan, D., Andriani, D.P., Das, R., Lau, A.: Investigating the impact of road condition complexity on driving workload based on subjective measurement using NASA TLX. MATEC Web Conf. EDP Sci. **136**, 02007 (2017)

8. Teh, E., Jamson, S., Carsten, O., Jamson, H.: Temporal fluctuations in driving demand: the effect of traffic complexity on subjective measures of workload and driving performance. Transp. Res. F: Traffic Psychol. Behav. **22**(1), 207–217 (2014)
9. Oviedo-Trespal Ac Ios, O., Hague, M.M., King, M., Washington, S.: Effects of road infrastructure and traffic complexity in speed adaptation behavior of distracted drivers. Accid. Anal. Prev. **101**(4), 67–77 (2017)
10. Ma, Y., Fan, L., Gu, G.: Risk identification of in-vehicle information system operation based on traffic environment complexity. In: Wang, W., Bengler, K., Jiang, X. (eds.) GITSS 2017. LNEE, vol. 503, pp. 845–855. Springer, Singapore (2019). https://doi.org/10.1007/978-981-13-0302-9_82
11. Horberry, T., Anderson, J., Regan, M.A., Triggs, T.J., Brown, J.: Driver distraction: the effects of concurrent in-vehicle tasks, road environment complexity and age on driving performance. Accid. Anal. Prev. **38**(1), 185–191 (2006)
12. Rudin-Brown, C.M., Edquist, J., Lenné, M.G.: Effects of driving experience and sensation-seeking on drivers' adaptation to road environment complexity. Saf. Sci. **62**, 121–129 (2014). https://doi.org/10.1016/j.ssci.2013.08.012
13. Xue, Q., Jiang, Y., Lu, J.: Risky driving behavior recognition based on trajectory data. China J. Highw. Transp. **33**(6), 84 (2020)
14. He, Q., Tu, H., Wei, H.: Analyzing car following driving ethogram based on driving simulator. J. Wuhan Univ. Technol. (Traffic Sci. Eng.). **45**(3), 420–423 (2021)
15. Li, S., Yin, Y., Wang, L., Xu, Y.: Quantification of driving fatigue based on entropy weight method. J. South China Univ. Technol. (Nat. Sci. Ed.). **45**(8), 50–56 (2017)
16. Hart, S., Staveland, J.: Development of NASA-TLX: results of empirical and theoretical research. Adv. Psychol. North-Holland. **52**, 139–183 (1988)
17. Li, L., Zhu, X., Ma, Z.: Driver brake reaction time under real traffic risk scenarios. Automot. Eng. **36**(10), 1225–1229 (2014)
18. Gao, K., Yang, Y., Li, A., Li, J., Yu, B.: Quantifying economic benefits from free-floating bike-sharing systems: a trip-level inference approach and city-scale analysis. Transp. Res. Part A: Policy Pract. **144**, 89–103 (2021). https://doi.org/10.1016/j.tra.2020.12.009
19. Liang, J., Feng, C., Song, P.: A survey on correlation analysis of big data. Chin. J. Comput. **39**(1), 1–18 (2016)

Driving Secondary Task Load Quantification Based on the AHP Algorithm Under the Voice Interaction Scenario

Wenlong Liu[1], Linhong Wang[1], and Ce Wang[2(✉)]

[1] Transportation College of Jilin University, Changchun 130022, Jilin, China
[2] Changchun FAWSN Automotive Technology Research and Development Co., Ltd., Changchun 130103, Jilin, China
cwang07@fawsn-autoparts.com

Abstract. With the rapid development of voice technology, Voice recognition technology has been bought into a large number of vehicle information systems, and different information from HMI displays during voice interaction will affect the driving status of drivers. In order to improve the interaction between the driver and the HMI (Human machine interaction) display to reduce driving distraction and to optimize the HMI information display, this study compared the difficulty of using the three common voice interaction interfaces based on voice interaction background and quantified the secondary task load generated by the voice interaction interface on the driver during driving. In this study, the indicators of head movement and the operation parameters of the vehicle during voice interaction are proposed to quantify the driving task load based on the hierarchical analysis method. To achieve the goal, 10 drivers are selected for driving simulation tests by using the UC-win/road driving simulator and head data acquisition software. During the test, every tester performs three-voice interaction tasks: map navigation, phone calls, and switching music. The results show that the proportion of driving load generated by the map navigation phone calling and switching music are 0.4898, 0.1992, and 0.311. Therefore, the HMI information display interface of map navigation and music switching needs to be simplified designed to reduce the presentation of redundant information. The study will provide a scientific basis for the voice interaction function of the HMI system and the information display design of the HMI interface.

Keywords: HMI character display · Voice interaction · Secondary task load · Hierarchical analysis · Interface information

1 Introduction

More frequent interactions between driver and vehicle-board information systems led to a high rate of traffic safety accidents 50% of Highway Administration NHTSA 2013 accidents are caused by the use of vehicle-board information systems when driving a vehicle [1]. Due to the high driving workload brought by the on-board voice task, the driver's attention to the main task is sometimes missing during the task time, and

Y. Bie et al. (Eds.): KES-STS 2022, SIST 304, pp. 11–21, 2022.
https://doi.org/10.1007/978-981-19-2813-0_2

cannot control the driving state of the vehicle in time [2]. Therefore, in the design and development of on-board information system, it is very necessary to determine the impact on driving safety when using common functions.

Actual road test: The driver performs a voice interaction task test on the actual road [3]. The biggest advantage of collecting the driver eye movement index and vehicle running state parameters from the voice interaction task during driving is the test results are real, but the cost and risk are high, and the reproducibility is poor, so the method can only conduct some simple test, in the more complex voice interaction task driving research, this method is generally not used [4].

Simple laboratory test: This kind of test refers to the simple laboratory configuration test without the simulation method [5], for example, using a computer to test traffic events on the computer screen. This has the advantages of simple cost and easy operation, but there is a large gap between the test and real driving, and the driving environment low the accuracy is low [6].

Simulation environment test: This kind of test uses a driving simulator with scenarios to simulate the real driving status. This method is adopted by most current studies, with the advantages of strong safety test design and easy test. Although the method is not as accurate as of the actual road test results, with the development of simulation technology, the driving simulator can give people the feeling of a real car [7]. And this kind of test can also reflect the actual driving status well, and it is easier to obtain all kinds of driver behavior and vehicle operation status parameters.

Considering the advantages and disadvantages of the above research methods, we choose the simulation environment test for the study.

Task 1: Control the on-board navigation device during driving. The Security Council of Canada, which in ref [8] is noted that in the process of destination information, the input information is less effective than the display of the information using sound control technology, the input or output information are transmitted through language, which can minimize the occupation of driver visual resources and reduce the impact on the driver's attention.

Task 2: Phone calling during driving: The task is mainly about the following conditions:1, telephone use type (contact and non-contact) 2, control method (sound control and touch control) 3, send receive messages and phone calling (different time length and different difficulty). Some previous research found that, in the above conditions during driving will damage impact the driver's reaction time and speed control, etc. Toernros et al. showed that phones are divided into handheld, and non-handheld phones, both non-handheld and handheld phones, which will have an impact on driving performance [9], The study of control mode by Ishida et al. showed that sound control mode can effectively reduce the impact of sub-task on driving performance [10]. Lesch et al. studied the use of the phone during driving, where the researchers asked the simple driver questions to answer the question accuracy as the evaluation criterion [11]. Horberry et al. found that the drivers need to think seriously or recall to record the content of the phone and send text messages, which will directly pose a threat to driving safety [12]. This study analyzed the effect of voice dialing on drivers.

Task 3: Listening to music during driving. The influence of such tasks on driving performance varies with the volume of music rhythm, and Schoemig derived from simulation trials of famous music students [13]. Different rhythms of music by taking up the driver attention space and increase the risk of traffic accidents, and the driver simulated driving speed, and speed estimates cause continuous impact within a certain range the faster the driver heart rate, the shorter the time the simulation driving time, the driver is prone to ignore the red light turn a blind eye to the zebra crossing driving error [14]. Oviedo's study showed that the higher the music decibel, the greater the stimulation to the driver, the greater the impact on driving performance [15]. The Mark study found that if the music rhythm is chosen properly and the volume is controlled at around decibels, it will help drivers relieve fatigue and shorten the response time in an emergency [16].

2 Voice-Based Interactive Driving Simulation Test

2.1 Design of Test

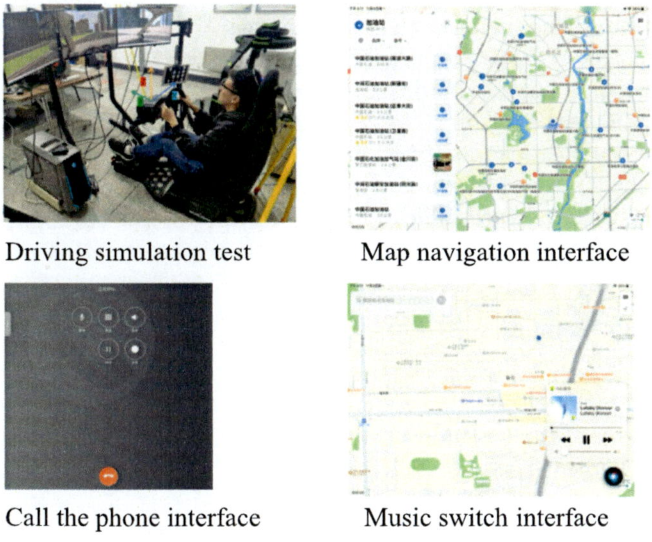

Fig. 1. Driving simulation test and voice interaction interface information display

The test is based on UC-win/road software and a driving simulator. The selected simulator scene is urban road, speed limit:70 km/h. The front car in the simulator is set driven freely. The following rear car in the simulator is controlled by the driver(tester) with the steering wheel and brake pedal. In the premise of ensuring driving safety, the diver can drive freely. The HMI monitor is simulated with a tablet computer. For the position of the HMI in the car, we assume the center of the steering wheel as the coordinate origin, and use a laser rangefinder of mm level to measure the relative position of the HMI monitor, and fix the HMI monitor to the corresponding position of the simulator according to the actual measurement. At the beginning of the test, the driver needs to wear the sensor device which is used to record head movement indicator, and then begins the driving simulation. During the driving simulation, the tester sends voice commands to the HMI display at any time to act as the sequence of 1, to search for the near gas station according to map navigation. 2, Phone calling; 3, Switching music. The tester must ensure safe driving when searching for the targets in a quiet test environment without any other interruption. And after the test, the tester must complete the preservation of the driving simulator and head movement data after the test. Ten drivers (mean = 24.5, standard deviation = 4.2) with driving experience are selected to conduct the driving simulation test (Fig. 1).

2.2 Test Index Collection

The relationship between head motion and the HMI display position is mainly reflected in the following two aspects, the first is the head motion time to search for the HMI task during driving, and the second is the amplitude of head motion. Head movement time determines whether the driver can drive safely, and the range of head movement determines the comfort of the driver's head.

The data of the driving simulator mainly reflects the driving performance of the driver. The head movement in the driving process affects the speed, the distance between the front and the rear, the time when the driver steps on the brake pedal, and the lateral control of the vehicle. And the simulator collects the data according to the above indicators: 1, the speed difference between the front and rear cars at the start and end of the head movement, 2, the distance between the front and rear cars at the start and end of the head movement, 3, the average value of the throttle opening and closing degree and 4, the lateral deviation of the vehicle during the head movement.

2.3 Test Data Processing

The obtained driving simulator data and head data were processed, and the results are shown in Fig. 2 and Fig. 3.

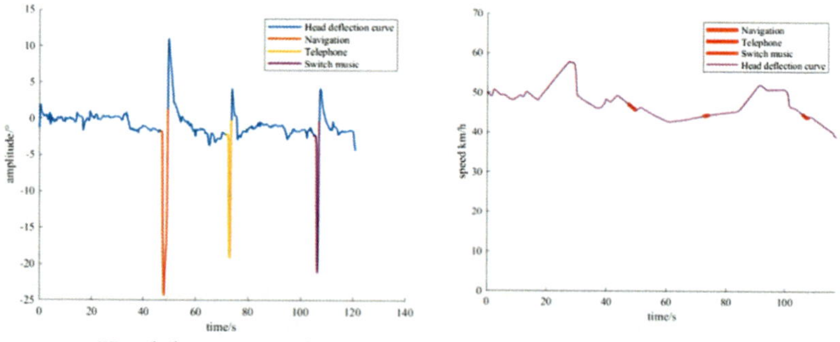

Head data processing Simulator data processing

Fig. 2. Head data matching and vehicle speed data matching

According to the head motion curve and driving simulator speed curve, the map navigation task of the driver has the largest amplitude and unstable speed control, the telephone task has the smallest amplitude and good vehicle control, the operation parameters of the music switching task are between the above two. The metrics extracted under the three Secondary tasks are shown in Tables 2, 3 and 4. Tables 2, 3 and 4 is to extract the driver's head deflection index and driving performance information (Table 1).

Table 1. Indicator encoding

Name of index	Code	Name of index	Code
Head movement time	HMT/s	Start distance	SD/m
Head dynamic amplitude	HDA/°	End distance	ED/m
Speed difference at the beginning	SDB/km·h^{-1}	Map navigation	MN
Speed difference at end	SDE/km·h^{-1}	Make a call	MC
Average opening and closing degree of accelerator pedal	AOCD/ω	Music switching	MS
Standard deviation of vehicle yaw	SDVY/°		

Table 2. Map navigation data extraction

Tester\Indicator	HMT	HDA	SDB	SDE	AOCD	SDVY	SD	SE
1	2.76	25.62	2.85	5.89	0.01	0.00	48.11	89.02
2	2.66	24.92	8.02	1.99	0.33	0.00	49.82	42.62
3	2.74	17.90	1.23	0.66	0.19	0.01	49.81	52.71
4	2.94	20.70	7.80	4.79	0.15	0.00	54.51	52.14
5	3.08	21.70	3.16	3.28	0.03	0.00	47.20	43.39
6	3.68	23.49	4.38	1.69	0.05	0.01	39.24	39.29
7	4.16	22.08	0.40	4.18	0.02	0.01	25.81	37.69
8	3.52	21.47	2.22	3.66	0.00	0.01	48.70	19.55
9	3.26	21.91	3.36	4.26	0.04	0.01	39.98	47.02
10	3.30	19.65	0.74	5.83	0.12	0.00	22.70	32.40

Table 3. Dial data extraction

Tester\Indicator	HMT	HDA	SDB	SDE	AOCD	SDVY	SD	SE
1	2.02	18.95	5.25	4.46	0.12	0.01	111.35	50.98
2	2.76	18.30	1.37	4.82	0.17	0.00	55.82	54.94
3	1.56	20.58	0.98	0.24	0.20	0.00	34.59	50.37

(*continued*)

Table 3. (*continued*)

Tester\Indicator	HMT	HDA	SDB	SDE	AOCD	SDVY	SD	SE
4	1.92	16.11	2.72	7.18	0.08	0.01	61.32	60.84
5	1.74	20.56	5.01	4.48	0.09	0.00	38.86	50.58
6	1.48	21.09	0.62	5.60	0.12	0.00	30.23	44.42
7	1.86	20.63	1.33	2.71	0.04	0.00	46.31	27.75
8	2.72	11.24	0.18	4.25	0.02	0.00	44.38	51.84
9	1.72	18.53	1.83	4.82	0.21	0.01	31.48	43.91
10	2.40	20.16	2.15	0.50	0.21	0.01	21.58	22.54

Table 4. Switch for music data extraction

Tester\Indicator	HMT	HDA	SDB	SDE	AOCD	SDVY	SD	SE
1	2.26	20.88	5.60	6.06	0.06	0.00	92.34	115.25
2	1.90	20.56	1.36	0.98	0.14	0.00	43.78	56.42
3	2.06	17.14	0.33	2.31	0.05	0.00	52.63	35.52
4	2.18	17.19	5.94	3.38	0.21	0.01	55.00	59.55
5	2.00	18.08	3.36	4.13	0.14	0.00	41.70	41.46
6	3.02	15.50	2.26	1.02	0.10	0.00	38.46	31.08
7	2.32	17.71	5.14	1.89	0.07	0.00	40.18	47.52
8	2.84	18.89	1.90	0.73	0.15	0.00	17.47	44.21
9	2.26	22.71	5.92	2.30	0.11	0.00	44.45	32.87
10	2.46	20.30	5.02	2.18	0.17	0.00	36.24	20.07

3 Model Construction

Analytic Hierarchy Process is the decision-making method that decomposes the elements related to decision-making into goals, guidelines, plans, etc., for qualitative and quantitative analysis. The proposed method has the advantages of systematization, flexibility, and simplicity. In this study, the degree of HMI display is evaluated by the head deflection index and driving performance index. The hierarchical analysis method meets the specific requirements of this study.

The process of creating a model via hierarchical analysis is as follows:

Step 1: Define the problem, determine the target;
Step 2: From the highest layer (target layer), through the middle layer (standard layer) to the lowest layer (scheme layer) to form a hierarchical structure model;
Step 3: Compare the scores to determine the score of the lower level to the upper level. Each standard layer does not necessarily have the same proportion with another. Each

of them has a certain proportion in the minds of decision makers. The scales 1–9 and their meanings are defined as Table 5: the judgment matrix.
Step 4: The hierarchical synthesis calculation;
Step 5: The consistency test.

Table 5. Judges the matrix scale definition

Scale	Meanings
1	Compared with the two indicators, they have the same importance
3	The former of the two indicators is slightly more important than the latter
5	The former is obviously more important than the latter
7	The former is strongly more important than the latter
9	The former is extremely more important than the latter
2, 4, 6, 8	The median of the above-mentioned adjacent judgment
Count backward	If the importance ratio of the index i to index j is a_{ij} then the importance ratio of the index j to the index i is $1/a_{ij}$

3.1 Build a Hierarchical Analysis Matrix

In order to analyze the difficulty of the interface of HMI display, the driver's workload is taken as the target layer. We selected eight standard layer indicators: head movement time, the head deflections, the speed difference of voice interaction, the standard deviation of the deviation, the mean of the vehicle accelerator pedal, the distance difference of voice interaction. Scheme layer selection: map navigation, making phone calls, and switching over music. According to the results of the test, the scale of each index is determined, and the scale matrix of the criterion layer and the significance layer is constructed (Fig. 3).

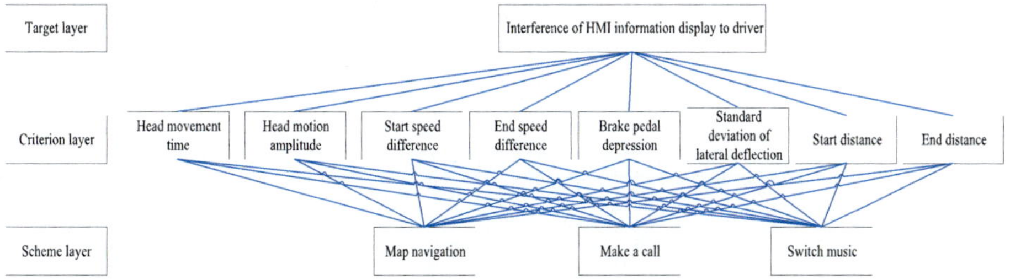

Fig. 3. Schematic diagram of the hierarchical analysis method structure

3.2 Calculate the Consistency Index CI

$$CI = \frac{\lambda_{max}}{n-1} \tag{1}$$

where *CI* is the consistency index, λ_{max} is the maximum eigenvalue of the judgment matrix, and *n* is the number of indicators at the level.

3.3 Find the Consistency Indicator RI

Table 6. Mean random consistency indicators

n	1	2	3	4	5	6	7	8	9
RI	0	0	0.52	0.89	1.12	1.24	1.36	1.41	1.46

where the RI in the table is the mean random consistency index (Table 6).

3.4 Calculate the Consistency Ratio, CR

$$CR = \frac{CI}{RI} \tag{2}$$

where *CR* is the consistency ratio, and when CR is less-than 0. 1, a one-time test is considered passed. Otherwise, some appropriate corrections should be made.

The second layer *CI* calculation and one-time test result is $\lambda_{\max} = 8.5841$, $CI = 0.0834$, $CR = 0.0592$, $CR < 0.10$. The third layer of index CI calculation and one-time test results are as shown in Table 10. All the metrics shown in Table 7 passed the consistency test. Table 8 scaling matrix was determined based on the Table 2, 3 and 4 driver performance data and Table 5. The third layer scaling matrix is obtained based on the mean value comparison of different voice interaction tasks. The main principle of comparison is: When the scale value is large, then the influence on the driver's driving workload is large. The third layer of scaling matrix was obtained based on the principles of the interface impact on drivers and Table 5, as shown in Table 9.

Table 7. Calculation and one-time test of the third layer of consistency index *CI*

Standard layer index	$\lambda_{\max}$	CI	CR	Pass?
HMT	3.0468	0.0234	0.0450	Yes
HDA	3.0536	0.0268	0.0516	Yes
SDB	3.0536	0.0268	0.0516	Yes
SDE	3.0536	0.0268	0.0516	Yes

(*continued*)

Table 7. *(continued)*

Standard layer index	λ_{max}	CI	CR	Pass?
AOCD	3.0055	0.0028	0.0053	Yes
SDVY	3.0536	0.0268	0.0516	Yes
SD	3.0536	0.0268	0.0516	Yes
ED	3.0536	0.0268	0.0516	Yes

Table 8. The second layer of the scaling matrix

	HMT	HDA	SDB	SDE	AOCD	SDVY	SD	ED
HMT	1.0	2.0	2.0	3.0	0.2	0.2	2.0	2.0
HDA	0.5	1.0	4.0	4.0	0.2	0.2	2.0	2.0
SDB	0.5	0.3	1.0	1.0	0.3	0.2	1.0	1.0
SDE	0.3	0.3	1.0	1.0	0.3	0.2	1.0	1.0
AOCD	5.0	5.0	4.0	4.0	1.0	0.3	5.0	5.0
SDVY	5.0	5.0	5.0	5.0	4.0	1.0	5.0	5.0
SD	0.5	0.5	1.0	1.0	0.2	0.2	1.0	1.0
ED	0.5	0.5	1.0	1.0	0.2	0.2	1.0	1.0

Table 9. The third layer of the scaling matrix(part)

Metric 1				Metric 2			
HMT	MN	MC	MS	HDA	MN	MC	MS
MN	1	3	3	MN	1	2	2
MC	0.33	1	0.5	MC	0.5	1	0.5
MS	0.33	2	1	MS	0.5	2	1

3.5 Weight Calculation

$$W_i = \frac{1}{n}\sum_{j=1}^{n} \frac{a_{ij}}{\sum_{k=1}^{n} a_{kj}}, i = 1, 2, \ldots, n \tag{3}$$

where W_i is the weight value, n is the number of metrics in the corresponding scaling matrix, and a_{ij} is the scaling value of column j, row i of the scaling matrix. Combining the scaling matrix Tables 8 and 9 and formula 3, the calculated degree of influence to different interfaces on the driver, as shown in Table 10.

According to the results of the hierarchical analysis method, the HMI interface of map navigation is the most difficult, with the largest Secondary task load to the driver. The

HMI information of the music playback interface is the least difficult, and the secondary task load to the driver is small. It shows that in the case of HMI voice interface design, the navigation interface design information is too much, and the driver's workload will increase when the driver acts voice interaction and observes the interaction effect. When optimizing the design, the unnecessary information display should be reduced to reduce the load of the driver voice interaction.

Table 10. Weight calculation of index level analysis

Weight	HMT	HDA	SDB	SDE	AOCD	SDVY	SD	SE	
	0.101	0.107	0.049	0.047	0.243	0.354	0.050	0.050	Weight
MN	0.590	0.491	0.312	0.198	0.539	0.491	0.491	0.491	0.490
MC	0.159	0.198	0.198	0.491	0.163	0.198	0.198	0.198	0.199
MS	0.252	0.312	0.491	0.312	0.297	0.312	0.312	0.312	0.311

4 Conclusions

We analyze the relationship between the head data of the testers and the operating parameter data of the simulated vehicle. The hierarchical analysis method is also used to quantify the driving load from the map navigation, phone call, and music switching interface of the voice interaction, and the scale judgment matrix is constructed according to the influence relationship between the head movement and the driving simulator. Finally, the Secondary task load results are calculated at different interfaces with the arithmetic average weight calculation method, and this makes us achieve the value of Secondary task load degree quantification. The calculation results show that among the three voice interaction interfaces, the map navigation interface is 0.4898, the caller is 0.1992, and switching music is 0.311. This indicates that the map navigation interface should be appropriately simplified to reduce the impact on the driver.

In later studies, other forms of voice interaction tasks or other forms of driving Secondary tasks can be analyzed, And further study can comprehensively improve the voice interaction system and reduce the workload generated by the voice interaction interface on drivers.

Acknowledgements. This study was supported by the National Key R&D Program of China under Grant (No. 2018YFB1600501), National Natural Science Foundation of China under Grant (No. 71971097), and Youth Program of National Natural Science Foundation of China (No. 52002143).

References

1. National Highway Traffic Safety Association (NHTSA). Traffic safety facts research note: distracted driving 2013. NHTSA, Washington (2013)

2. Gressai, M., Varga, B., Tettamanti, T., Varga, I.: Investigating the impacts of urban speed limit reduction through microscopic traffic simulation. Commun. Transp. Res. **1**, 100018 (2021). https://doi.org/10.1016/j.commtr.2021.100018
3. Pushpa, C., Nagendra, R.V.: Modelling driver distraction effects due to mobile phone use on reaction time. Transp. Res. Part C Emerg. Technol. **77**, 351–365 (2017)
4. Jazayeri, A., Martinez, J.R.B., Loeb, H.S., Yang, C.C.: The Impact of driver distraction and secondary tasks with and without other co-occurring driving behaviors on the level of road traffic crashes. Accid. Anal. Prev. **153**, 106010 (2021). https://doi.org/10.1016/j.aap.2021.106010
5. Ismaeel, R., Hibberd, D., Carsten, O., Jamson, S.: Do drivers self-regulate their engagement in secondary tasks at intersections? An examination based on naturalistic driving data. Accid. Anal. Prev. **137**, 105464 (2020). https://doi.org/10.1016/j.aap.2020.105464
6. Wolf, P., Rausch, J., Hennes, N., Potthast, W.: The effects of joint angle variability and different driving load scenarios on maximum muscle activity – a driving posture simulation study. Int. J. Ind. Ergon. **84**, 103161 (2021). https://doi.org/10.1016/j.ergon.2021.103161
7. Xiong, Z., Olstam, J.: Orchestration of driving simulator scenarios based on dynamic actor preparation and automated action planning. Transportation Research Part C: Emerg Technol **56**, 120–131 (2015). https://doi.org/10.1016/j.trc.2015.02.008
8. Merat, N., Jamson, A.H., Lai, F.C.H., Carsten, O.: Highly automated driving, secondary task performance, and driver state. Hum. Factors **54**(5), 762–771 (2012)
9. Törnros, J.E.B., Bolling, A.K.: Mobile phone use—effects of handheld and handsfree phones on driving performance. Accid. Anal. Prev. **37**(5), 902–909 (2005). https://doi.org/10.1016/j.aap.2005.04.007
10. Qi, W., Shen, B., Yang, Y., Qu, X.: Modeling drivers' scrambling behavior in China: an application of theory of planned behavior. Travel Behav. Soci. **24**, 164–171 (2021). https://doi.org/10.1016/j.tbs.2021.03.008
11. Lesch, M.F., Hancock, P.A.: Driving performance during concurrent cell-phone use: are drivers aware of their performance decrements? Accid. Anal. Prev. **36**(3), 471–480 (2004). https://doi.org/10.1016/S0001-4575(03)00042-3
12. Horberry, T., Anderson, J., Regan, M.A., Triggs, T.J., Brown, J.: Driver distraction: the effects of concurrent in-vehicle tasks, road environment complexity and age on driving performance. Accid. Anal. Prev. **38**(1), 185–191 (2006). https://doi.org/10.1016/j.aap.2005.09.007
13. Schömig, N., Metz, B., Krüger, H.-P.: Anticipatory and control processes in the interaction with secondary tasks while driving. Transp. Res. F Traffic Psychol. Behav. **14**(6), 525–538 (2011). https://doi.org/10.1016/j.trf.2011.06.006
14. Xu, Y., Zheng, Y., Yang, Y.: On the movement simulations of electric vehicles: a behavioral model-based approach. Appl. Energy **283**, 116356 (2021). https://doi.org/10.1016/j.apenergy.2020.116356
15. Oviedo-Trespalacios, O.: Getting away with texting: Behavioural adaptation of drivers engaging in visual-manual tasks while driving. Transp. Res. Part A Policy Pract. **116**, 112–121 (2018). https://doi.org/10.1016/j.tra.2018.05.006
16. Vollrath, M., Schleicher, S., Gelau, C.: The influence of cruise control and adaptive cruise control on driving behaviour – a driving simulator study. Accid. Anal. Preven. **43**(3), 1134–1139 (2011). https://doi.org/10.1016/j.aap.2010.12.023

advanced driving assistance systems (ADAS) such as forward collision warning (FCW), automatic emergency braking (AEB), adaptive cruise control (ACC), and lane departure warning (LDW) [2–4]. These systems are used to guide driving behavior in real-time, thus preventing human error, leading to accidents. ADAS measures the position relationship between the vehicle and the road system in real-time through onboard sensors. When a specific dynamic parameter of the vehicle exceeds the warning threshold, the system will provide the driver with sound, light, or tactile warning information feedback in time [5]. Since ADAS could promptly remind drivers to take evasive measures such as deceleration or steering, most ADAS users believe that reliable driving assistance systems can effectively improve driving safety. However, in practical applications, when the FCW predicts a collision or NCEs, different drivers have different evasive behaviors [6].

Due to the difficulty of obtaining large amounts of collision data, the crash surrogate is widely used in the research of driving risk. NCEs are common crash surrogates, which are defined as a combination of steering, braking, decelerating, and control inputs [7]. Near-crash data can provide controllable laboratory data as a useful supplement to traffic safety research [8]. Time headway (THW), time to collision (TTC), TTC^{-1}, longitudinal deceleration, operation response time, steering wheel angle, etc., have been extensively and quantitatively adopted to study car-following behaviors to replace traditional safety indicators [9]. However, taking THW and TTC as general indicators for ADAS assessment and risk quantification still have limitations [10]. THW only retains vehicle speed information without considering the influence of relative speed, while TTC only retains relative speed information without considering the impact of vehicle speed. Besides, the expected averages of THW and TTC between drivers with different characteristics are also very different [11]. Drivers' evasive behaviors are affected by many factors, such as Drivers' expectations, gender, age, cognitive load, and the urgency of driving conditions [12].

Given the above, many efforts had been made to explorer the influence of ADAS on driving behavior and to evaluate the effectiveness and reliability of ADAS. However, most studies focused on the impact of ADAS on driving risk and the acceptance of different types of drivers under normal driving conditions. Few studies pay attention to the influence of ADAS on drivers' evasive behavior under dangerous driving conditions. Thus, this study focused on exploring the influence of ADAS on drivers' evasive behaviors during car-following in NCE on highways. An on-road experiment was conducted on public highways in real traffic. The dangerous driving data of drivers in NCE under different working states of ADAS (i.e., activated or not) were collected. K-means clustering method was used to classify the risk level of NCE. The effects of the risk level of NCE, the working state of ADAS and THW on the probability of drivers choosing evasive behavior are analyzed. The research results will help to objectively and real-time evaluate the impact of ADAS on highways car-following and evasive behaviors and better enhance the ability of safe driving assistance.

2 Methodology

The analyses of this study had three main parts: (1) Combined with maximum longitudinal deceleration, average longitudinal deceleration, and percentage of vehicle kinetic energy reduction, the risk levels of NCEs were divided into different groups by using K-means clustering. (2) The probability differences of drivers in different risk groups choosing different evasive behaviors were studied. (3) The internal influence of ADAS engagement on different evasive behaviors was evaluated by the THW index.

2.1 Experiment

To investigate the effectiveness of the ADAS, we recruited 20 drivers to participate in the field operation test, including six female drivers and 14 male drivers. Their average age is 32 (ranging from 26 to 53) years old. Their practical driving experience was more than five years. All experimental vehicles were required to be equipped with a forward collision warning device based on the Beidou navigation satellite system (equivalent to ADAS with only FCW function). The test route was located at Xinbo Expressway in Guangdong Province, China, as shown in Fig. 1. The test route had a length of 25 km, a design speed of 120 km/h, and six lanes in both directions. In the test, the test vehicles were required to run in clusters, and the speed is maintained at 80–100 km/h.

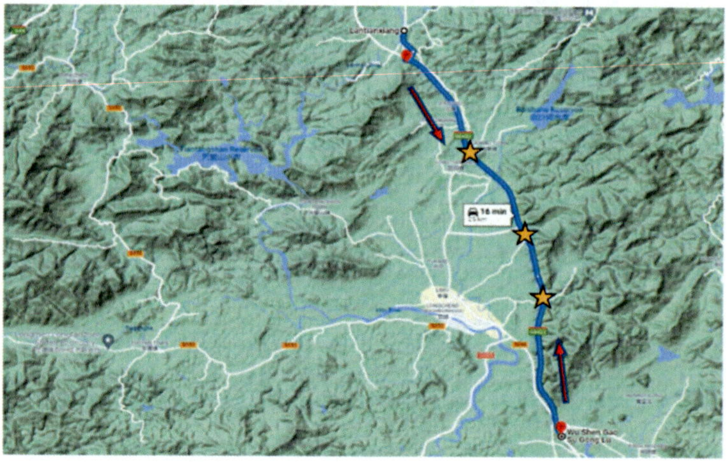

Fig. 1. The test route (blue line), showing the begin/end point (red dot to up), the northern turn-around point at Lantian Interchange (red dot to down), and the three points at which participants were asked to do dangerous driving tasks while driving (three yellow five-pointed star points along route)

2.2 Data Extraction

The raw data collected in this study and the preprocessed data types were presented in Table 1. The original data included the real-time vehicle position data collected through the onboard terminal and the highway alignment design data collected in advance. The preprocessed data included data related to the mapping relationship between vehicles and roads and the interaction relationship between vehicles and vehicles.

Table 1. Raw data and preprocessed data list

Raw data		Preprocessed data	
Vehicle terminal	Highway	Vehicle-road mapping relationship	Vehicle-to-vehicle interaction
Vehicle ID Time (*ms*) Coordinate (*X, Y*) Speed (*km/h*) Engagement state	Highway Alignment Design data	Lateral distance (*m*) Lane location (1, 2, 3) Accumulated mileage (*m*)	Preceding vehicle's ID Acceleration (*m/s2*) Spatial headway (*m*) Time headway (*s*)

2.3 Modeling Evasive Behaviors

The evasive behaviors of NCEs in this study consisted of two types: car-following and overtaking. Based on the differences in the interaction between the front and rear vehicles in the two different evasive behaviors, this study carried out quantitative modeling and descriptions.

(1) Car-following
The two test vehicles with the smallest difference in mileage in the same lane could be defined as the front and rear vehicles. The car-following state could be determined when the THW of the front and rear vehicles was less than 5 s [13]. The Gazis-Herman-Rothery (GHR) model was used as the car-following model [14], as shown in Eq. (1).

$$a_n = \alpha V_n(t)^\beta \frac{\Delta V_n(t-\tau_n)}{\Delta X_n(t-\tau_n)^\gamma} \tag{1}$$

where, V_n is the speed of the rear vehicle (*km/h*), a_n is the acceleration of the rear vehicle (m/s2); ΔV_n is the relative speed of the front and rear vehicles (km/h); ΔX_n is the distance between the front and rear vehicles (m), τ_n is reaction time (s); n is the nth vehicle (i.e. the rear vehicle); α, β, γ as parameters.

(2) Overtaking
A minimum safety distance (MSD) model for lane changing [15] was used to judge the safe lane-changing behavior of test vehicle m at any time t, as shown in the Eq. (2).

$$y_n(t) - y_m(t) > \frac{L_m}{2}cos\theta(t) + \frac{W_m}{2}sin\theta(t) + \frac{L_n}{2} \tag{2}$$

where, L_m denotes the length of the vehicle m; W_m denotes the width of the vehicle m; L_n denotes the length of the vehicle n; θ is the angle between the tangent direction of the track of vehicle m and the lane line.

Based on the mathematical models of different evasive behaviors, this study proposed the judgment rules for Drivers' emergency evasive behaviors in NCEs, as shown in Table 2.

Table 2. Two typical hedging behavior judgment rules

Evasive behaviors	Judgment rules
Car-following	• The headway value was less than 5s • Minimum headway of the front and rear vehicles • The position and speed relationship with the preceding vehicle satisfied Eq. (1)
Overtaking	• The headway value was less than 5s • The position and speed relationship with the preceding vehicle satisfied Eq. (2)

2.4 K-means Clustering

Cluster analysis can classify driving risks involved in different NCEs into different risk levels and had been used to assess individual driver risks [16]. In previous studies, maximum longitudinal deceleration, average deceleration, and vehicle kinetic energy reduction ratio were widely used to classify the driving risk levels of NCEs [10]. This study used the following characteristics to characterize the driving risk level involved in typical NCEs during naturalistic driving:

(1) maximum longitudinal deceleration d_{max}.
(2) The average deceleration d_{ave}, it was calculated as follows:

$$d_{ave} = \frac{1}{t_1 - t_0}\int_{t_0}^{t_1} a(t)dt = \frac{v(t_1) - v(t_0)}{t_1 - t_0} \tag{3}$$

where, $v(t)$ and $a(t)$ denote the vehicle's velocity and acceleration; t_0 is the initial moment when longitudinal deceleration is greater than 2.0 m/s^2; t_1 is the corresponding moment of maximum deceleration.

(3) The percentage reduction in vehicle kinetic energy η_r, it was calculated as follows:

$$\eta_r = \frac{0.5 \times mv(t_0)^2 - 0.5 \times mv(t_1)^2}{0.5 \times mv(t_0)^2} = 1 - \left(\frac{v(t_1)}{v(t_0)}\right)^2 \tag{4}$$

where m denotes the vehicle mass.

K-means clustering is an iterative algorithm for clustering analysis. K-means clustering divides the data into k groups in advance and randomly selects k objects as the initial cluster centers. Finally, according to the distance between each object and seed cluster center, each object is assigned to the cluster center closest to it [17]. This process can be described as follows:

$$\underset{\omega}{\operatorname{argmin}} \sum\nolimits_{i=1}^{k} \sum\nolimits_{x \epsilon \omega_i} \|x - \mu_i\|^2 \tag{5}$$

where x is the observed data, $\omega = [\omega_1, \omega_2, \ldots, \omega_n]$ denotes the set of k clusters and μ_i denotes the mean point of cluster set ω_n.

3 Results

3.1 NCEs Classification by K-means Clustering

In the end, a total of 347 groups of NCEs that met the criteria were obtained, of which 185 groups were without warning, and 163 groups were with a warning. This study used the K-means clustering algorithm to divide NCEs into three groups: high, medium, and low risk. The output of the cluster analysis was shown in Fig. 2. Table 3 summarized the statistical characteristics of the three driving risk groups. The classification results shown that: 1) The distribution of NCEs in different risk groups followed a pyramid structure. This result was consistent with previous similar research results [18]. 2) The average longitudinal deceleration of the high-risk group was about twice that of the low-risk group. 3) ADAS engagement could effectively reduce the proportion of medium and high-risk NCEs by more than 10%.

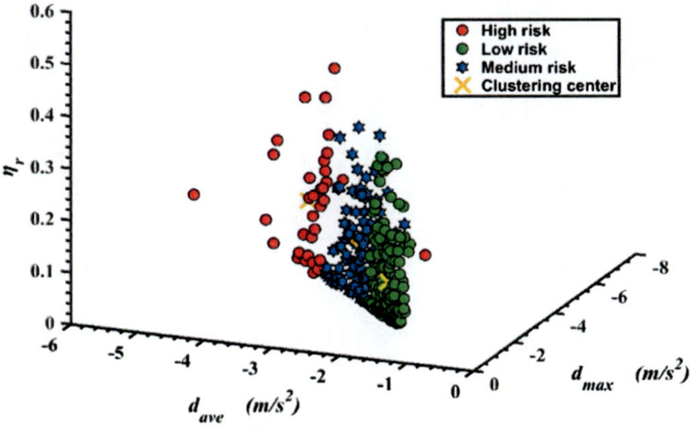

Fig. 2. Clustering results of driving risk under different ADAS states

Table 3. Risk classification of NCEs in various states of ADAS

Risk groups	Number of NCEs			Percentage (%)			Cluster center feature mean		
	ADAS ON	ADAS OFF	Total	ADAS ON	ADAS OFF	Δ	$d_{\max}$	d_{ave}	η_r
Low risk	92	109	201	46	54	8	−1.969	−1.873	0.088
Medium risk	49	60	109	45	55	10	−2.721	−2.535	0.123
High risk	15	6	37	41	59	18	−3.908	−3.643	0.163

3.2 Comparison of the Probability of Different Evasive Behaviors

As the results illustrated in Fig. 3, the probability of drivers choosing car-following in low, medium, and high-risk NCEs was greater than overtaking. For the low-risk group, ADAS engagement could increase the probability that drivers chose overtaking behavior, making the probability that drivers chose car-following (P = 55%) and overtaking (P = 45%) were almost equal. ADAS engagement could reduce the probability that drivers chose overtaking for the medium and high-risk groups. More than 70% of drivers were more willing to slow down and follow the vehicle to avoid danger.

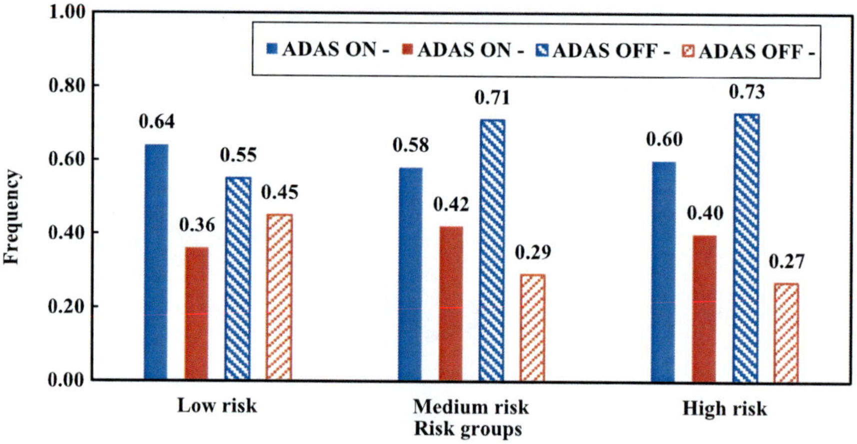

Fig. 3. Probability of different evasive behaviors for different risk groups

3.3 Analysis of the Influence of THW on Different Evasive Behaviors

The Shapiro-Wilk test first showed that the THW from car-following and overtaking behaviors did not obey the normal distribution ($p < 0.05$, rejecting the null hypothesis of the normal distribution), which violated the normal hypothesis of the T-test and ANOVA. Thus, the Mann-Whitney U test was used to check the significant influence of ADAS engagement on the headway of different risk-averse behaviors. The Mann-Whitney U test is a non-parametric alternative to the T-test. When the normal distribution and the homogeneity of variance do not meet the requirements of the T-test, the Mann-Whitney U test should be used [19]. According to the Mann-Whitney U test results shown in Table 4, ADAS engagement showed significant differences in the THW of overtaking behaviors ($p < 0.05$).

Table 4. THW statistical test results of different evasive behaviors in ADAS OFF and ON states

Evasive behaviors	Sample size	Mean	S.D	Mann-whitney U test
Car-following	114	1.395	0.931	ADAS OFF vs. ADAS ON: $p = 0.412$
Overtaking	131	1.376	1.056	ADAS OFF vs. ADAS ON: $p = 0.038$

As shown in Fig. 4(a-b) and Fig. 4(d-e), when ADAS was turned on, the THW of the two evasive behaviors of car-following and overtaking presents an overall increasing trend. Figure 4(c) and (f) were the box-type comparison diagrams of the THW of different evasive behaviors. The results showed that when ADAS was turned on, the average THW of car-following increased from 1.32 s to 1.46 s (an increase of 0.14 s), while the average THW of overtaking increased from 1.13 to 1.59 s (an increase of 0.46 s). Thus, ADAS engagement had a greater impact on THW when overtaking.

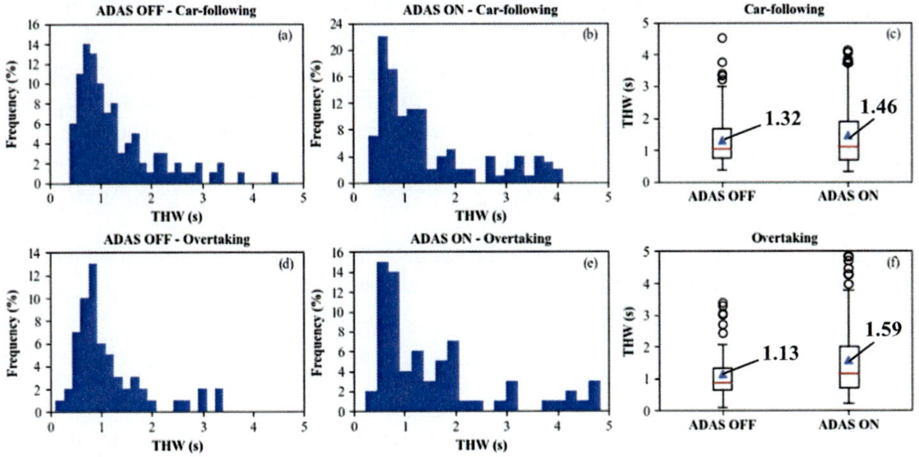

Fig. 4. Boxplot of minimum THW corresponding to following and overtaking behavior of ADAS OFF and ON states. The frequency distribution of THW was depicted using the data recorded from the following behavior and overtaking behavior.

4 Discussion and Conclusion

Cluster analysis is an effective method to classify the driving risks in different NCEs into different levels [20]. Consistent with other studies, the optimal number of clusters for driving risk in NCEs was 3 [17, 21]. The driving risks were classified into three categories (low, medium, and high risk) using the K-means clustering method combining maximum longitudinal deceleration, average deceleration, and vehicle kinetic energy reduction ratio. The above three indicators indicated that speed and deceleration during emergency braking were significant risk variables.

Results of this study showed that ADAS could effectively reduce the proportion of NCEs, and it also affected the driver's choice of evasive behaviors. ADAS engagement could reduce the ratio of medium and high-risk NCEs by more than 10%. For the low-risk group, ADAS engagement could increase the probability that drivers choose overtaking behavior. For the medium and high-risk groups, ADAS engagement could reduce the probability that drivers choose overtaking, and more than 70% of drivers were more willing to choose car-following. ADAS engagement could improve the driver's risk perception and early response capabilities. The study also found that ADAS engagement can help drivers maintain a greater headway with the vehicle ahead during the avoidance process. ADAS engagement had a significant impact on the THW of overtaking behavior. When ADAS was turned on, the average THW for car-following behavior increased from 1.32 s to 1.46 s, while the average THW for overtaking behavior increased from 1.13 to 1.59 s.

Conducting a real vehicle test that simulated a critical highway crash event was very dangerous, which might lead to some limitations of this research. To ensure the safety of this test, the participants were all 25~40 years old drivers of a passenger transport company with skilled driving skills, and the test was carried out under clear weather conditions. The study results ignored the influence of gender, age, driving skills, weather, and traffic flow on Drivers' evasive behaviors. The driving risk varied greatly among individual drivers [22]. Despite the study's limitations, the methods and results of this study still evaluated the impact of ADAS on highway aversion behaviors in an objective and reliable way.

Acknowledgment. This project was jointly sponsored by the National Natural Science Foundation of China (52102416) and Natural Science Foundation of Shanghai (22ZR1466000).

References

1. Yu, B., Bao, S., Zhang, Y., Sullivan, J., et al.: Measurement and prediction of driver trust in automated vehicle technologies: an application of hand position transition probability matrix. Transp. Res. Part C: Emerg. Technol. **124**(0968–090X), 102957 (2021)
2. Gao, K., Tu, H., Sun, L., Sze, N.N., Song, Z., Shi, H.: Impacts of reduced visibility under hazy weather condition on collision risk and car-following behavior: implications for traffic control and management. Int. J. Sustain. Transp. **14**(8), 635–642 (2020). https://doi.org/10.1080/15568318.2019.1597226
3. Yu, B., Bao, S., Feng, F., Sayer, J.: Examination and prediction of drivers' reaction when provided with V2I communication-based intersection maneuver strategies. Transp. Res. Part C: Emerg. Technol. **106**, 17–28 (2019)
4. Chen, Y., Xie, X., Bo, Y., Li, Y., Lin, K.: Multitarget vehicle tracking and motion state estimation using a novel driving environment perception system of intelligent vehicles. J. Adv. Transp. **2021**, 1–16 (2021). https://doi.org/10.1155/2021/6251399
5. Bao, S., Wu, L., Yu, B., Sayer, J.R.: An examination of teen drivers' car-following behavior under naturalistic driving conditions: with and without an advanced driving assistance system. Accid. Anal. Prev. **147**, 105762 (2020)
6. Gao, K., Yang, Y., Sun, L., Qu, X.: Revealing psychological inertia in mode shift behavior and its quantitative influences on commuting trips. Transp. Res. F: Traffic Psychol. Behav. **71**, 272–287 (2020). https://doi.org/10.1016/j.trf.2020.04.006

7. Yu, B., Bao, S., Chen, Y., LeBlanc, D.J.: Effects of an integrated collision warning system on risk compensation behavior: an examination under naturalistic driving conditions. Accid. Anal. Prev. **163**, 106450 (2021)
8. Dozza, M.: What factors influence drivers' response time for evasive maneuvers in real traffic? Accid. Anal. Prev. **58**, 299–308 (2013)
9. Pan, C., Xu, J., Fu, J.: Effect of gender and personality characteristics on the speed tendency based on advanced driving assistance system (ADAS) evaluation. J. Intell. Connected Veh. **4**(1), 28–37 (2021). https://doi.org/10.1108/JICV-04-2020-0003
10. Xu, Y., Zheng, Y., Yang, Y.: On the movement simulations of electric vehicles: a behavioral model-based approach. Appl. Energy **283**, 116356 (2021). https://doi.org/10.1016/j.apenergy.2020.116356
11. Xu, J., Shu, H.-B., Shao, Y.-M.: Modeling of driver behavior on trajectory–speed decision making in minor traffic roadways with complex features. IEEE Trans. Intell. Transp. Syst. **20**(1), 41–53 (2018)
12. Wang, J., Zheng, Y., Li, X., Yu, C., et al.: Driving risk assessment using near-crash database through data mining of tree-based model. Accid. Anal. Prev. **84**, 54–64 (2015)
13. Ali, Y., Zheng, Z., Haque, M.M.: Modelling lane-changing execution behaviour in a connected environment: a grouped random parameters with heterogeneity-in-means approach. Commun. Transp. Res. **1**, 100009 (2021). https://doi.org/10.1016/j.commtr.2021.100009
14. Yue, L., Abdel-Aty, M., Wang, Z.: Effects of connected and autonomous vehicle merging behavior on mainline human-driven vehicle. J. Intell. Connected Veh. **5**(1), 36–45 (2022). https://doi.org/10.1108/JICV-08-2021-0013
15. Qi, W., Shen, B., Yang, Y., Qu, X.: Modeling drivers' scrambling behavior in China: an application of theory of planned behavior. Travel Behav. Soc. **24**, 164–171 (2021). https://doi.org/10.1016/j.tbs.2021.03.008
16. Wu, K.F., Aguero-Valverde, J., Jovanis, P.P.: Using naturalistic driving data to explore the association between traffic safety-related events and crash risk at driver level. Accid. Anal. Prev. **72**, 210–218 (2014)
17. Guo, F., Simons-Morton, B.G., Klauer, S.E., Ouimet, M.C., et al.: Variability in crash and near-crash risk among novice teenage drivers: a naturalistic study. J. Pediatr. **163**(6), 1670–1676 (2013)
18. Kodinariya, T.M., Makwana, P.R.: Review on determining number of cluster in k-means clustering. Int. J. **1**(6), 90–95 (2013)
19. McKnight, P.E., Najab, J.: Mann-Whitney U test. In: Weiner, I.B., Craighead, W.E. (eds.) The Corsini Encyclopedia of psychology. p. 1, Wiley, Hoboken (2010)
20. Lyu, N., Duan, Z., Ma, C., Wu, C.: Safety margins–a novel approach from risk homeostasis theory for evaluating the impact of advanced driver assistance systems on driving behavior in near-crash events. J. Intell. Transp. Syst. **25**(1), 93–106 (2021)
21. Zheng, Y., Wang, J., Li, X., Yu, C., et al.: Driving risk assessment using cluster analysis based on naturalistic driving data. In: 17th International IEEE Conference on Intelligent Transportation Systems (ITSC), pp. 2584–2589 (2014)
22. Xu, Y., Ye, Z., Wang, C.: Modeling commercial vehicle drivers' acceptance of advanced driving assistance system (ADAS). J. Intell. Connected Veh. **4**(3), 125–135 (2021). https://doi.org/10.1108/JICV-07-2021-0011

Effect Factors Analysis of Driver's Freeway Route Deviation Based on Questionnaire Survey Data

Nanjie Zhou[1], Huapeng Wang[2], Wenyi Wang[2], and Weiwei Qi[2](✉)

[1] Guangzhou Northring Intelligent Transport Technology Co., Ltd., Guangzhou 510030, China
[2] School of Civil Engineering and Transportation, South China University of Technology, Guangzhou 510641, China
ctwwqi@scut.edu.cn

Abstract. In order to investigate the effect factors of route deviation when drives use navigation application on freeways, we conducted a route deviation questionnaire on 219 drivers. The driver cognitive patterns were summarized from the questionnaire by exploratory factor analysis, and effective explanatory variables were screened by chi-square test. Meanwhile, related social factors were screened by chi-square test and Tukey's Honestly Significant Difference (HSD) test. Finally, a binary logistic regression model was established to reveal the influence weight of related factors on deviation from the route. The results show that navigation application design, navigation usage habits, traffic environment and education can influence drivers' use of navigation, with navigation application design having the greatest degree of influence, followed by navigation usage habits, and then traffic environment interference and education. Specifically, navigation application that is consistent with most drivers' habits can reduce the probability of route deviation; those with good driving habits have less route deviation; complex traffic environments can increase the probability of route deviation; the probability of route deviation decreases as education increases. The research contributes to the optimization of freeway navigation application.

Keywords: Route deviation · Freeway · Navigation language optimization · Logistic regression

1 Introduction

In recent years, navigation map application has developed rapidly as an essential tool for travel planning and route navigation. Although the overall performance of navigation maps tends to improve the efficiency of drivers' travel in unfamiliar environments and solve congested sections [1, 2], the partial inability of existing navigation systems still has optimization potential. Even, some studies found that it has some negative effects on driving behavior, for example, Tamer et al. concluded that it can distract drivers with the help of navigation systems, which may lead to an increase in road accidents, and investigated that navigation system display size, ambient lighting, and gender can affect

Y. Bie et al. (Eds.): KES-STS 2022, SIST 304, pp. 32–43, 2022.
https://doi.org/10.1007/978-981-19-2813-0_4

driving safety [3, 4]; Kaber analyzed that when the driver's vision and cognition are occupied by navigation information, it will increase the response time and the number of operation errors [5]; Jamson et al. found that both visual and auditory navigation tasks delayed the driver's response to unexpected conditions [6]. Thus, navigation still has optimization potential, e.g., Zhao identified the effect of variable message signs on individual path selection behavior by studying the different effects on different groups [7]; Lee explored the best strategy for delivering navigation information to drivers by controlling the type and mode of information [8]; Larsson proposes pro-social control strategies for vehicles that take into account driving comfort and traffic efficiency [9]; Li propose a heuristic routing algorithm to identify the feasible routing paths for shared rides that interest both ridesharing drivers and riders [10]. Shi explored factors related to autonomous driving safety [11].

In terms of the navigation voice prompt method, Uang et al. noted that the choice of command or descriptive voice announcements depending on the content of the information was beneficial in improving driver compliance with prompts [12]; Large et al. showed that high trust speech increased driver trust in navigation but did not significantly affect driver attention to wayfinding signs [13]; Lavie et al. found that drivers performed best when using the least informative map [14]. The above studies reveal the effects of navigation information giving methods on driving behavior and subjective cognition, and point the way for further optimization of navigation information giving methods. In terms of speech wording, Dalton et al. found that simple voice commands were easier to follow and less intrusive to drivers than complex voice commands [15]; Rasker et al. suggested that navigation voice prompts don't need to provide road names because it is not easily and quickly understood by the driver [16]. Visually, Lin et al. found no significant difference in driving performance between 2D and 3D map, while sweeping behavior was more frequent in 3D than 2D, and drivers made significantly fewer navigation errors when using the sub-window navigation display [17].

As for drivers, Yang explored the factors that influence drivers' willingness to use mobile navigation applications and found that the influencing factors are attitude, perceived usefulness, driver orientation, navigation application affinity, and perception of distraction [18]. Meanwhile, Bian's study found that an interaction between prompt timing and prompt message of the voice navigation system affect the driver's psychological state and vehicle operation on urban highways [19]; Ali and Fu did some research on lane change conditions separately [20, 21]; Imants believes that a deeper understanding of how drivers use multiple sources of traffic information can help improve driver safety and comfort, increase the availability of information sources, and help reduce driver stress, anxiety and information overload while driving [22]. Pan explored the relationship between speed behavior of participants and driving styles [23]. In addition, Emmerson has found that the use of in-car navigation systems provides better road information for older drivers, and that further improvements in navigation design are needed to improve the quality of service for older drivers [24].

From the above research, it is clear that there is still potential for optimizing navigational speech. In this paper, the influencing factors of route deviation will be studied and analyzed from the driver's cognitive pattern, hoping to explore the navigation language that is more in line with the driver's cognitive habits, make the electronic navigation system better assist the driver on the freeway.

2 Research Content

2.1 Questionnaire Design

Combined with the existing questionnaires [18], the questionnaire of this study is designed based on the problems of voice prompts in navigation application and the actual situation of drivers using navigation application, in order to investigate the factors influencing drivers' deviation from the route when using navigation on freeways. The questionnaire consists of two parts: the first part is a non-scale questionnaire, which collects basic information about the respondents, such as gender, age and whether they often deviate from the route when using navigation on freeway; the second part is a scale questionnaire, with 17 questions related to the use and design of navigation (Table 1). The scale questions were scored on a 5-point Likert scale: 1 - strongly disagree, 2 - disagree, 3 - barely agree, 4 - relatively agree, 5 - strongly agree.

Table 1. Question content

Question	Content
Q6	Trust the navigation can provide useful information
Q7	When you go off course, it's mostly a navigation issue
Q8	Voice prompt is too quick to think and judge accordingly
Q9	Voice prompt is too late, resulting in delayed response
Q10	Voice prompt is too early and are disruptive to current driving
Q11	The default voice is too monotonous to focus on the prompt
Q12	Insufficient voice prompt times result in poor prompt effect
Q13	More focus on navigation voice than screen
Q14	I will take action immediately when hearing the prompt
Q15	I slow down and notice the traffic conditions when hearing the prompt
Q16	Focus on navigation information such as routes, turns, and distances
Q17	Focus on location information such as maps, locations, and driveways
Q18	Focus on violation information such as speed limit
Q19	Focus on road conditions information such as traffic congestion,
Q20	There are more traffic around, which affects my operation
Q21	The prompt doesn't match the road markings
Q22	The prompt doesn't match the actual road

2.2 Respondents

The survey was focused on drivers in the age of 20 to 64 given that drivers aged 18–20 are still in their driving practice period and have little experience in freeway driving, as well as the minority of drivers over 65 who use mobile phones while driving. This questionnaire was distributed in April 2021. To ensure that the surveyed drivers fill in the questionnaire as honestly as possible, a statement is made at the beginning of the questionnaire to inform the respondents of the purpose and meaning of this questionnaire. In the end, 231 questionnaires were returned with 12 questionnaires with missing information, 219 questionnaires were valid, with an efficiency rate of 94.8%.

The basic information statistics of the survey objects are shown in Table 2. Among them, 107 were males and 112 were females with similar percentages. The age was divided into segments by every 15 years, with 40.18% of the survey objects aged 20–34, 40.64% aged 35–49 and 19.18% aged 50–64. The percentage of driving experience less than 3 years was 29.22% (including 11.41% less than 1 year; 17.81% 1–3 years) and greater than 3 years was 70.78%. Education was divided according to junior high school and below, high school, bachelor and postgraduate, accounting for 12.79%, 36.53%, 42.92%, and 7.76%, respectively.

Table 2. Information of participants

Information	Categories	Number	Proportion
Gender	Male	107	48.86%
	Female	112	51.14%
Age	20–34	88	40.18%
	35–49	89	40.64%
	50–64	42	19.18%
Driving experience	Less than 1 year	25	11.41%
	1–3 years	39	17.81%
	More than 3 years	155	70.78%
Education	Junior high and below	28	12.79%
	High school	80	36.53%
	Bachelor	94	42.92%
	Postgraduate	17	7.76%

3 Effect Factors of Route Deviation

3.1 Exploratory Factor Analysis (EFA)

An exploratory factor analysis was conducted on the 17 questions that reflect the use of navigation application, to determine the corresponding factor structure. The KMO test and Bartlett's spherical test were conducted on the questionnaire data: the KMO measure was 0.857, indicating that the sample size was suitable for factor analysis; the Bartlett's spherical test result was 2002.298 (the significance level $P = 0.000 < 0.001$), indicating the possibility of common factors among the observed variables. Then, the results of the Scree Test are shown in Fig. 1, the first 3 triangles are above the bend of the curve, indicating that there are 3 factors with eigenvalues greater than 1, so the questionnaire data are suitable for extracting 3 common factors.

The factor structure and its loadings are shown in Table 3: Factor 1, named as navigation application design, contains 7 questions, mainly reflecting the driver's feelings about navigation and the reasonableness of navigation voice settings, with a variance contribution rate of 26.476%; Factor 2, navigation usage habits, contains 7 questions, mainly reflecting the driver's behavioral habits and concerns about navigation information, with a variance contribution rate of 25.631%; Factor 3 is traffic environment interference, containing 3 questions, reflecting the influence of traffic environment on navigation use, with a variance contribution rate of 13.815%, and a cumulative variance contribution rate of 65.922% for the 3 common factors.

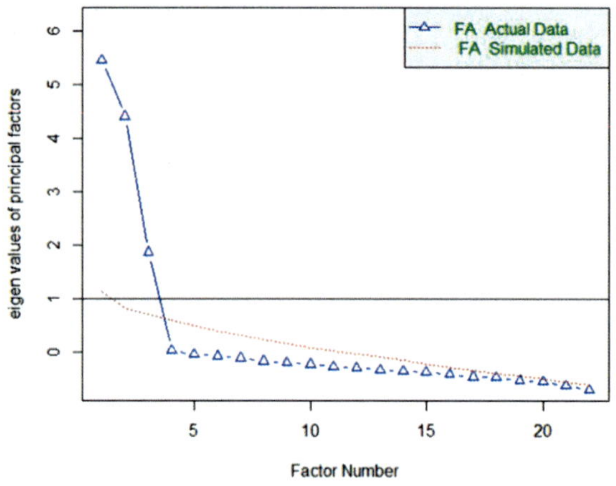

Fig. 1. Parallel analysis scree plots

Table 3. The factor loading matrix

Question	Navigation design	Usage habits	Traffic environment
Q6	0.733		
Q7	0.813		
Q8	0.805		
Q9	0.779		

(continued)

Table 3. (*continued*)

Question	Navigation design	Usage habits	Traffic environment
Q10	0.774		
Q11	0.757		
Q12	0.812		
Q13		0.761	
Q14		0.808	
Q15		0.795	
Q16		0.793	
Q17		0.811	
Q18		0.849	
Q19		0.778	
Q20			0.853
Q21			0.875
Q22			0.870

3.2 Questionnaire Reliability and Validity Tests

Cronbach's alpha reliability coefficient method was used to test the internal and overall reliability of the questionnaire. The test results showed that the alpha coefficient of navigation application design, navigation usage habits, traffic environment interference and the overall questionnaire were 0.896, 0.906, 0.845, and 0.841, which were higher than 0.80, indicating that the internal and overall questionnaire had good reliability. Meanwhile, the cumulative variance contribution of the three public factors according to the exploratory factor analysis was 65.922%, which is higher than 60%, indicating that the questionnaire has good structural validity.

4 The Effect of Driver Cognitive Models on Route Deviation

One-way ANOVA was used to analyze the differences in driver cognitive patterns and whether route deviations occurred frequently, and the results are shown in Table 4. Navigation application design, navigation usage habits, and traffic environment interference were significantly different in whether or not to deviate ($P < 0.01$). The 219 samples were divided into two small samples based on whether they deviated from the route (153 for the frequently deviated sample and 66 for the infrequent deviated sample). The mean and variance of each question score were respectively calculated for the two samples, while the chi-square test was used to further determine whether there was a significant difference between each question, and the results are shown in Table 5.

It is apparent that the frequently deviated group score higher than the infrequently deviated group on each question, and the scores are distributed around 4 (relatively agree), indicating that drivers who use navigation and frequently deviate are more dependent on the navigation application, while their driving behavior is more influenced by navigation. In the results of the chi-square test, each question of the 3 factors has a significant difference for frequent route deviation, with the navigation application design factor has a more significant difference for frequent route deviation. Therefore, for improving the driver's behavior of route deviation when using navigation application on the freeway, we should first consider the navigation application's own design.

Table 4. Analysis of variance

Variable	Navigation design		Usage habits		Traffic environment	
	F-value	Saliency	F-value	Saliency	F-value	Saliency
Route deviation	61.534	0.000**	19.472	0.000**	22.562	0.000**

Note: ** indicates $P < 0.01$; * indicates $P < 0.05$

Table 5. Chi-square test

Question		Mean (Standard deviation)		Results
		Frequently deviate	Infrequently deviate	
Factor 1:	Q6	4.16(0.97)	3.20(1.40)	0.000
	Q7	4.03(1.06)	3.12(1.23)	0.000
	Q8	3.96(1.09)	2.89(1.22)	0.000
	Q9	4.07(1.00)	2.91(1.29)	0.000
	Q10	4.01(1.05)	2.79(1.29)	0.000
	Q11	4.08(1.00)	3.23(1.30)	0.000
	Q12	4.06(1.04)	2.95(1.30)	0.000
Factor 2:	Q13	3.91(1.18)	3.14(1.16)	0.000
	Q14	3.97(1.04)	3.27(1.25)	0.000
	Q15	4.05(1.10)	3.33(1.35)	0.002
	Q16	3.99(1.06)	3.32(1.39)	0.003
	Q17	3.94(1.14)	3.52(1.19)	0.008
	Q18	3.86(1.12)	3.30(1.24)	0.001
	Q19	3.94(1.06)	3.21(1.26)	0.000
Factor 3:	Q20	4.11(1.03)	3.48(1.04)	0.000
	Q21	3.99(1.04)	3.03(1.12)	0.000
	Q22	4.08(0.96)	3.33(1.26)	0.000

5 The Effect of Demographic Factors on Route Deviation

Gender, age, driving age and education were subjected to a chi-square test to explore the effect on route deviation, and the results are shown in Table 6. Education was associated with whether route deviation occurred frequently ($P = 0.000 < 0.01$). The participants with high school degrees accounted for 45% of frequent route deviations; a bachelor's degree accounted for 64% of infrequent route deviations. To further investigate differences in education levels, a multiple comparison analysis of the two was conducted using the Tukey's HSD test, and the results are shown in Fig. 2. There is a significant difference between education below high school and education above the bachelor level at the confidence interval 95% level, indicating that the higher the education level, the less frequent route deviation occurs when using navigation application, which confirms that the level of education is a factor in driving accidents. Based on the above tests, high school can be used as a cut-off to redefine the classification and values of the education variable: 1 - high school and below, 2 - bachelor's degree or higher.

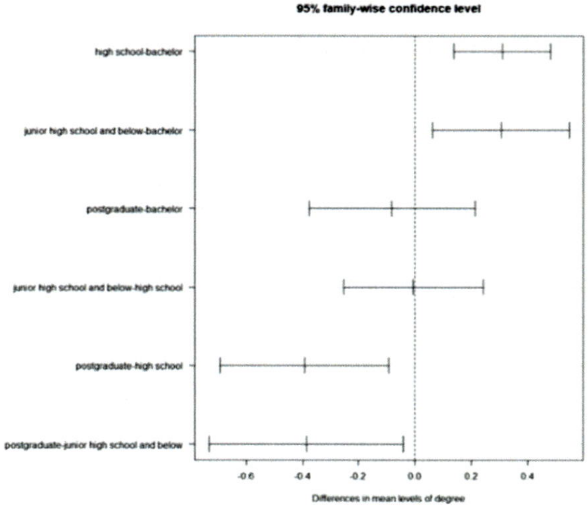

Fig. 2. Multiple comparison of means-Tukey HSD

6 Response to Effect Factors of Route Deviation

To understand the extent of which navigation application design, navigation usage habits, traffic environment interference, and educational attainment could affect deviation, a binary logistic regression model is established to make predictions. Whether a driver frequently occurs route deviation is a dichotomous quantity, where 0 indicates infrequent or no deviation and 1 indicates frequent deviation. The model uses whether route deviation occurs frequently as the dependent variable, and uses the three factors and the education variable as the independent variables, to establish a model for freeway route deviation. The results obtained are shown in Table 7. The significance level of each variable is less than 0.01 and can be used to predict the likelihood of route deviation. The Hosmer-Lemeshow test value for the overall model was 2.318 ($P = 0.970 > 0.05$), indicating that the model fits well and the prediction accuracy is 85.40%. Therefore, the four factors can be used as inputs to the route deviation prediction model.

According to the regression coefficients of the independent variables, the design of the navigation application has the greatest influence on deviate from the route. Specifically, the more the design of the navigation voice and display screen conforms to the cognitive habits of most drivers, the less deviations from the route occur. The second is the driver's habit of using navigation. Those who have good driving habits can get information from the navigation tips that are beneficial to them to reduce the deviation from the route. Furthermore, the traffic environment can reduce the effectiveness of navigation prompts and increase the probability of route deviation. Finally, the higher the driver's education, the lower the probability of route deviation. The model is useful for predicting route deviation when drivers use navigation application on the freeway, improving navigation prompts and reducing the probability of route deviation.

Table 6. Chi-square test for demographic factors

Information	Categories	Proportion		Results
		Frequently deviate	Infrequently deviate	
Gender	Male	46%	55%	0.269
	Female	54%	45%	
Age	20–34	36%	50%	0.127
	35–49	44%	32%	
	50–64	20%	18%	
Driving experience	Less than 1 year	10%	15%	0.518
	1–3 years	18%	17%	
	More than 3 years	72%	68%	
Education	Junior high and below	16%	6%	0.000**
	High school	45%	17%	
	Bachelor	34%	64%	
	Postgraduate	5%	13%	

Table 7. Logistics regression analysis results

Independent variable	B	Wald	Sig	Exp(B)
Education	−2.170	20.038	0.000**	0.114
Navigation application design	1.695	39.809	0.000**	5.449
Navigation usage habits	1.328	27.759	0.000**	3.775
Transportation environment	1.106	20.932	0.000**	3.021
Constant	5.114	31.802	0.000**	166.368

Note: ** indicates $P < 0.01$; * indicates $P < 0.05$

7 Conclusion

In this paper, a questionnaire was used to analyze the use of navigation applications by 219 drivers, and the concept of the cognitive model was introduced to illustrate the factors influencing the occurrence of route deviation when drivers use navigation. In general, navigation application that meets most drivers' usage habits can reduce the probability of route deviation; those with good driving habits have fewer cases of deviations from the route; and complex traffic environments can enhance the probability of route deviation. The paper also proposes corresponding optimization suggestions for navigation application based on navigation application design factors.

Social factors also influence the use of navigation by drivers, specifically, the probability of route deviation decreases with increasing education. In addition, the negative correlation between education and the impact of the traffic environment suggests that the key point for a better traffic environment is the development of traffic awareness and quality of traffic participants, so that to improve traffic safety in society, drivers need to be systematically educated and trained, as well as social propaganda.

Acknowledgments. This work was supported by the National Natural Science Foundation of China (NO.52072131), the Key Research Projects of Universities in Guangdong Province (NO.2019KZDXM009), the Science and Technology Project of Guangzhou City (NO.201804010466), and the Special Innovative Projects of Universities in Guangdong Province (NO.2019GKTSCX036).

References

1. Liu, P., Liu, Y.: Optimal information provision at bottleneck equilibrium with risk-averse travelers. Transp. Res. Rec. **2672**(48), 69–78 (2018)
2. Liu, Y., Nie, Y.: A credit-based congestion management scheme in general two-mode networks with multiclass users. Netw. Spat. Econ. **17**(3), 681–711 (2017). https://doi.org/10.1007/s11067-017-9340-7
3. Yared, T., Patterson, P., All, E.S.A.: Are safety and performance affected by navigation system display size, environmental illumination, and gender when driving in both urban and rural areas? Accid. Anal. Prev. **142**, 105585 (2020)

4. Yared, T., Patterson, P.: The impact of navigation system display size and environmental illumination on young driver mental workload. Transp. Res. F: Traffic Psychol. Behav. **74**, 330–344 (2020). https://doi.org/10.1016/j.trf.2020.08.027
5. Kaber, D.B., Liang, Y., Zhang, Y., Rogers, M.L., Gangakhedkar, S.: Driver performance effects of simultaneous visual and cognitive distraction and adaptation behavior. Transp. Res. F: Traffic Psychol. Behav. **15**(5), 491–501 (2012)
6. Hamish Jamson, A., Merat, N.: Surrogate in-vehicle information systems and driver behaviour: effects of visual and cognitive load in simulated rural driving. Transp. Res. F: Traffic Psychol. Behav. **8**(2), 79–96 (2005)
7. Zhao, W., Ma, Z., Yang, K., Huang, H., Monsuur, F., Lee, J.: Impacts of variable message signs on en-route route choice behavior. Transp. Res. Part A: Policy Pract. **139**, 335–349 (2020). https://doi.org/10.1016/j.tra.2020.07.015
8. Bumho, L., Yoo Jin, N.L., Sanghoo, P., Hyunsik, K., Su-Jin, L., Jinwoo, K.: Driver's distraction and understandability (eou) change due to the level of abstractness and modality of gps navigation information during driving. Procedia Comput. Sci. **39**, 115–122 (2014)
9. Larsson, J., Keskin, M.F., Peng, B., Kulcsár, B., Wymeersch, H.: Pro-social control of connected automated vehicles in mixed-autonomy multi-lane highway traffic. Commun. Transp. Res. **1**, 100019 (2021). https://doi.org/10.1016/j.commtr.2021.100019
10. Li, Y., Liu, Y., Xie, J.: A path-based equilibrium model for ridesharing matching. Transp. Res. Part B: Methodol. **138**, 373–405 (2020)
11. Gao, K., Yang, Y., Qu, X.: Diverging effects of subjective prospect values of uncertain time and money. Commun. Transp. Res. **1**, 100007 (2021). https://doi.org/10.1016/j.commtr.2021.100007
12. Uang, S.T., Hwang, S.L.: Effects on driving behavior of congestion information and of scale of in-vehicle navigation systems. Transp. Res. Part C: Emerg. Technol. **11**(6), 423–438 (2003). https://doi.org/10.1016/S0968-090X(03)00003-2
13. Large, D.R., Burnett, G.E.: The effect of different navigation voices on trust and attention while using in-vehicle navigation systems. J. Saf. Res. **49**(69), e61-75 (2014)
14. Lavie, T., Oron-Gilad, T., Meyer, J.: Aesthetics and usability of in-vehicle navigation displays. Int. J. Hum Comput. Stud. **69**(1), 80–99 (2011)
15. Dalton, P., Agarwal, P., Fraenkel, N., Baichoo, J., Masry, A.: Driving with navigational instructions: investigating user behaviour and performance. Accid. Anal. Prev. **50**, 298–303 (2013)
16. Rasker, P.C., Post, W.M., Schraagen, J.M.C.: Effects of two types of intra-team feedback on developing a shared mental model in command & control teams. Ergonomics **43**(8), 1167–1189
17. Ching-Torng, L., Hsin-Chieh, W., Ting-Yen, C.: Effects of e-map format and sub-windows on driving performance and glance behavior when using an in-vehicle navigation system. Int. J. Ind. Ergon. **40**(3), 330–336 (2010)
18. Liping, Y., Yang, B., Xiaohua, Z., Xiaoming, L., Xianglin, Y.: Drivers' acceptance of mobile navigation applications: an extended technology acceptance model considering drivers' sense of direction, navigation application affinity and distraction perception. Int. J. Hum. Comput. Stud. **145**, 102507 (2021)
19. Yang, B., Xiaolong, Z., Yiping, W., Xiaohua, Z., Hao, L., Yuelong, S.: Influence of prompt timing and messages of an audio navigation system on driver behavior on an urban expressway with five exits. Accid. Anal. Prev. **157**, 106155 (2021)
20. Ali, Y., Zheng, Z., Haque, M.M.: Modelling lane-changing execution behaviour in a connected environment: a grouped random parameters with heterogeneity-in-means approach. Commun. Transp. Res. **1**, 100009 (2021). https://doi.org/10.1016/j.commtr.2021.100009
21. Fu, C., Sayed, T.: Bayesian dynamic extreme value modeling for conflict-based real-time safety analysis. Anal. Methods Accid. Res. **34**, 100204 (2022)

22. Qi, W., Shen, B., Yang, Y., Qu, X.: Modeling drivers' scrambling behavior in China: an application of theory of planned behavior. Travel Behav. Soc. **24**, 164–171 (2021). https://doi.org/10.1016/j.tbs.2021.03.008
23. Pan, C., Xu, J., Fu, J.: Effect of gender and personality characteristics on the speed tendency based on advanced driving assistance system (ADAS) evaluation. J. Intell. Connected Veh. **4**(1), 28–37 (2021). https://doi.org/10.1108/JICV-04-2020-0003
24. Xu, Y., Ye, Z., Wang, C.: Modeling commercial vehicle drivers' acceptance of advanced driving assistance system (ADAS). J. Intell. Connected Veh. **4**(3), 125–135 (2021). https://doi.org/10.1108/JICV-07-2021-0011

Demand Analysis of Customizable Car Sharing Functions Based on Kano Model

Daming Li[1], Hongyu Ren[2], Shuolei Qin[2], Quan Yuan[3], and Weiwei Qi[2](✉)

[1] Guangzhou Northring Intelligent Transport Technology Co., Ltd., Guangzhou 510030, China
[2] School of Civil Engineering and Transportation, South China University of Technology, Guangzhou 510641, China
ctwwqi@scut.edu.cn
[3] State Key Laboratory of Automotive Safety and Energy, School of Vehicle and Mobility, Tsinghua University, Beijing 100084, China

Abstract. As people's demand for travel increases, sharing travel, especially car-sharing, is receiving more and more attention and anticipation. In order to fully study the characteristics of customizable car sharing products and the demand tendency of consumers, on the basis of determining the advantages of customizable car sharing, this paper intends to explore and analyze the personalized needs of users of customizable car sharing products through a series of qualitative and quantitative research methods. A great deal of research has been carried out by scholars at home and abroad on the analysis of consumer needs and satisfaction, creating a good theoretical basis and reference for the research work in this paper. From the data obtained from the questionnaire survey, this paper summarized the customizable attribute items of product and their customization priorities and proposed an analysis method of personalized demand items based on the Kano model. Firstly, the Kano model combined with the fuzzy clustering method is used to identify and screen the personalized demand items of 24 initial car sharing products obtained from the survey, and the hierarchy of demand model of the product is constructed. Then, according to the personalized demand items screened by the Kano model, the questionnaire is designed for the importance survey. Based on the data obtained from the importance survey, the initial weights of the underlying personalized demand items of the hierarchy of demand model are calculated, and the entropy method is used to adjust the weights. After that, the importance ranking of 12 personalized demand items of car sharing products is determined. The results show that the security attribute has the highest overall importance among the 12 customized functions. This fully shows that the safety customization function of car sharing is most valued by consumers, and its demand and expectation are strong. The research results can provide certain guiding significance for customized production of car sharing companies.

Keywords: Car sharing · Fuzzy clustering · Quantitative Kano model · Hierarchy of needs

Y. Bie et al. (Eds.): KES-STS 2022, SIST 304, pp. 44–54, 2022.
https://doi.org/10.1007/978-981-19-2813-0_5

1 Introduction

1.1 Background and Study Motivation

With the increase of people's travel demand, there exists a serious imbalance between the development of road and energy resources and the growth of private car ownership. For one thing, there is a huge potential scale of consumption in the car market. For another, there is a huge disparity between the current population and car ownership. At the same time, given recent incontrovertible evidence that the environment (and our planet) will face serious threats unless we humans finally act quickly, some are calling for a tougher stance against the misuse of private cars [1]. Under severe traffic and environmental pressures, shared mobility is receiving more and more attention and expectation.

When car sharing companies provide customization services, the richer the selection of customizable attributes offered, the more detailed the customer's individual needs will be. The more customer needs are met, the higher the satisfaction generated. At the same time, however, the cost of production, the complexity of the production process, and the difficulty of maintenance increase as the number of customizable attribute items increases. Due to cost and capacity constraints, it is not possible for car sharing companies to use all customizable attribute items, so only a select number of core items are open for customization.

To this end, this paper will use a personalized demand acquisition method based on the Kano model to analyze the personalized demand characteristics of consumers from their point of view. Furthermore, the customizable attribute items of the car sharing and their customization priorities are presented in a graphical format.

1.2 Current Research Status

Advances in digital technology will further develop the sharing economy. In the field of transportation, autonomous vehicle technology will accelerate the development of customizable car sharing, and will change the way of land use and traffic system composition. Akimoto used a global energy system model to quantify the impact of customizable travel and car sharing on global energy demand and emissions reductions [2]. Wang' s research on some shared electric transportation products shows that when the average utilization rate of shared cars is low, the energy loss of batteries in an idle state is large. So improving the utilization of shared cars can reduce energy waste [3]. Lu [4] proposed a cooperative adaptive cruise control method for the mixed traffic flow of autonomous vehicles and manned vehicles. This method can reduce fuel consumption and improve the environmental friendliness of the transportation system. The research results of Kumakoshi [5] show that the influence of self-driving car sharing on land use and transportation is heterogeneous, which highlights the importance of formulating appropriate urban and transportation planning strategies. Wu [6] proposed a shared automatic rail transport system. The system can adapt to dynamic traffic demand and is not limited by fixed routes and timetables. Studies have shown that this automated transportation system will not increase energy consumption while shortening travel time.

In a study of the behavior and attitudes of riders and car-sharing practitioners, Xu [7] used the random forest method to evaluate the influencing factors of commercial vehicle drivers' acceptance of the advanced driver assistance system (ADAS). The results show that duration time, vehicle speed, and driver age are three key factors. Chen [8] combined the protection motivation theory with the use situation theory to explore the mechanism of passenger protection behavior in the environment of car sharing and the findings can help enterprises of car sharing to take appropriate strategies to improve safety. Etminani-Ghasrodasht [9] discussed the concerns and preferences of passengers, especially disabled passengers, after integrating self-driving shared buses into existing carpooling services. Hartl [10] uses the social problem framework to analyze the contribution of car sharing to the needs of different groups.

Krauss [11] uses a mixed logit model, considering four shared travel modes and three traditional travel modes, to calculate the respective travel times as well as entrance, exit and parking search times. At the same time, part-worth analysis is used to analyze the importance of individual attributes.

In 1984, Kano created a Kano model based on customer experience and product objective performance. Fatma [12] proposed to integrate the Kano model into the architecture case to clarify the quality of architecture design. Xu [13] established an analytical Kano (A-Kano) model based on customer demand analysis for the inherent defects of the traditional Kano method. The model can effectively integrate customer preferences into product design, and achieve the optimal balance between customer satisfaction and producer capacity.

The interpretation of personalized needs from a narrow perspective can be directly regarded as customized needs, that is, enterprises set a scope limit of customization and design for customers to choose. Ma [14] used perceived value theory, trust theory, and transaction cost theory to construct a structural equation model to explain passenger loyalty. The results show that enterprises should provide differentiated services for passengers according to the purpose of car sharing. Based on the concept of "customer customization", Wind [15] proposes to help customers identify and define their unique needs based on various cutting-edge Internet technologies. Tseng [16] indicated that parameter setting and classification tree could be used to feedback personalized customer needs, and improved the method of selecting established products.

2 Research Contents and Methods

2.1 Interview Research and Questionnaire Survey

Interview survey and identification of the initial personalized demand items of products: design a series of questions about the habits and preferences of online ride-hailing drivers and passengers, and conduct qualitative research through individual interviews or collective conversations.

Firstly, relevant information and literature such as industry information reports are consulted, and user needs are combed in major information platforms. At the same time, combined with personal experience of sharing automotive product services, sorting out the product quality characteristics that may be customized. On this basis, the demand of consumers is qualitatively analyzed. According to the preliminary user needs, in-depth interviews with users of shared automobile products and obtain the detailed needs of the group for customized shared automobiles. Then fully tap the consumer demand for customized shared car models and functions, and provide research direction for Kano questionnaire design.

In the actual survey, eight in-depth interviewees were selected. Including 3 drivers and 5 passengers. Through in-depth interviews with these eight objects, a total of 30 original needs items were sorted out.

Later, after the initial personalized demand affinity graph model was obtained through a small range of pre-survey, the Kano questionnaire was designed around the items at the bottom of the initial model. Kano questionnaire in all the functional attributes are open two questions, one positive and one negative, combined with five options, details can refer to Table 1.

Table 1. Kano two-factor questionnaire

Question: how do you feel	I like it	It should be	I don't care	I can stand	I hate it
If it has a function?	○	○	○	○	○
If it doesn't have a function?	○	○	○	○	○

The Kano questionnaire is delivered by random sampling, and the sample data are collected by electronic questionnaire. The target group is users of car sharing (including drivers and general customers). A total of 300 questionnaires were delivered, and 233 valid questionnaires were finally recovered. The effective rate of the questionnaire was 77.6%.

2.2 Survey Method of Importance Degree

This article is based on the hierarchical model to obtain the bottom of the requirements. In order to further understand and clarify the importance of these needs, special questionnaires are prepared to investigate and clarify the importance. The specific form of questionnaire is the question of scale ten, which is based on a standard importance questionnaire. This survey must be carried out after personalized needs screening.

3 Building a Hierarchy of Needs Model

3.1 Identify and Screen the Kano Category of Personalized Demand Items for Car Sharing

Interviews can be used to identify the initial personalized demand items for the product: Set up a series of questions about the habits and preferences of drivers and passengers, and conduct qualitative research through individual interviews or group conversations. The initial personalized demand items are shown in Table 3.

Referring to the survey results, the membership degree of each demand item under each Kano category is calculated by comparing and analyzing the Kano evaluation table. In the Kano evaluation table, according to the impact on satisfaction when user needs are met, user needs are classified into five categories. The comparison table is shown in Table 2

Table 2. Comparison table of Kano model evaluation results

Functions/Services		Negative questions				
		I hate it	I can stand	I don't care	It should be	I like it
Positive questions	I hate it	Q	R	R	R	R
	I can stand	M	I	I	I	R
	I don't care	M	I	I	I	R
	It should be	M	I	I	I	R
	I like it	O	A	A	A	Q

A: Charm attribute; O: Expected attribute; M: Necessary attribute; I: No difference attribute; R: Reverse attribute; Q: Suspicious attribute.

Table 3. Coding of initial personalized demand items

Initial personalized demand items		
1. Side airbags	2. Vehicle data recorder	3. Panoramic images
4. Emergency call for accident	5. Automatic parking	6. Parking assist
7. Adaptive cruise control	8. Lane keeping system	9. Behavior detection system
10. AED Emergency System	11. Air cleaning system	12. Car roof display
13. Automotive connectivity ecosystem	14. Single side electric sliding door	15. Intelligent lighting system
16. Intelligent windshield	17. Smart NFC key	18. Multifunction chair
19. New warehouse	20. In-car Wi-Fi	21. Central control system
22. Catering facilities	23. Personal entertainment facilities	24. Takeout/Express delivery system

First, re-encode the personalized demand items, details are shown in Table 3.

The Kano category membership degree of each personalized demand item is obtained from the Kano questionnaire data, and the maximum membership principle is used to judge.

However, when the different classification membership degrees of the same personalized demand item are similar, the mixed classification judgment standard proposed by Lee and Newcomb is adopted. That is, the values of TS (Total Strength) and CS (Category Strength) are calculated. When TS ≥ 0.6 and CS ≤ 0.06, it is a mixed class attribute. Otherwise, the Kano classification of the demand item is directly determined by the 'maximum membership principle'. The calculation methods of TS and CS are as follows:

$$TS = \frac{A + O + M}{A + O + M + Q + R + I} \tag{1}$$

$$CS = \frac{\text{Maximum} - \text{Minimum}}{A + O + M + Q + R + I} \tag{2}$$

Based on the above membership principle, the following table is used to determine whether the personalized demand item belongs to a mixed classification (Table 4).

Table 4. Kano questionnaire survey results

Functions	3	4	5	6	7	9	11
Classification	M, I	A, I	I, O	O, I	O, I	O, I	O, I
TS	0.672	0.642	0.622	0.682	0.622	0.662	0.667
CS	0.01	0.05	0.03	0.06	0	0.06	0.025
Functions	13	15	16	17	19	20	
Classification	O, I	I, M	I, O	O, I	I, O	O, I	
TS	0.622	0.657	0.622	0.662	0.617	0.652	
CS	0.03	0.045	0.035	0.04	0.04	0.015	

By observing the above tables, it is not difficult to find that the demand items 3,4,5,6,7,9,11,13,15,16,17,19,20 conform to the characteristics of 'TS $\geq$ 0.6 and CS $\leq$ 0.06'. They have two sets of similar membership values, so they are classified as 'mixed ', meaning that the Kano category to which they belong cannot be directly determined. Therefore, it is necessary to introduce fuzzy clustering in the end to classify the personalized demand items and clarify their Kano categories. The detailed process is as follows:

① According to the five categories of Kano and the corresponding personalized needs, create a fuzzy similar matrix. The items distributed in the lowest level of the model are set to X_i, and the number of demand items is n. A five-dimensional vector $(X_{i1}, X_{i2}, X_{i3}, X_{i4}, X_{i5})$ is constructed for any X_i, where the demand items corresponding to X_{i1}~X_{i5} correspond to the membership degrees of M, O, A, I and R respectively. Thus, we can get the fuzzy similarity matrix R, where r_{ij} is the similarity of X_i and X_j. Using the absolute value subtraction method, r_{ij} can be obtained:

$$r_{ij} = \begin{cases} 1, i = j \\ 1 - c\sum_{k=1}^{n} \left|X_{ik} - X_{jk}\right|, x \geq 0 \end{cases} (i, j = 1, 2, \cdots n) \tag{3}$$

Parameter c is a correction factor, valued at (0, 1).$0 \leq r_{ij} \leq l$ 。

② Solving transfer closure $t(R)$. The minimum fuzzy equivalent matrix $t(R)$ containing the similar matrix R is constructed, and the $t(R)$ is obtained by the square method:

$$R^2 = R \cdot R = \bigvee_{n}^{k=1} \left(r_{ik} \bigwedge r_{jk}\right) \tag{4}$$

In the above expression, $\vee$ denotes the maximum value from two elements, and $\wedge$ denotes the minimum value. After a certain number of operations, there must be a positive integer m, so that $R^{2m} = R^m$, and $t(R) = R^{2m}$.

③ Clustering analysis and generating clustering diagram. Since $t(R)$ is a fuzzy equivalent matrix, when $\lambda \in [0, 1]$, the truncation matrix R_λ is an ordinary equivalent matrix, which can be used to divide X. The calculation method of R_λ is as follows: when $r_{ij} \geq \lambda$, let $r_{ij} = 1$; when $r_{ij} < \lambda$, let $r_{ij} = 0$. Thus, $R_\lambda = (r_{ij})n \times n$ is calculated. When the submatrix elements constructed by relative rows and columns are 1, they can be classified as the same clustering. When Λ changes from 1 to 0, this means classifying from fine to coarse, integrating in turn, and finally getting the dynamic clustering graph. Then the corresponding λ value is selected on-demand, and the corresponding R_λ is combined to clarify the Kano type of the requirement class and the personalized requirement. When Λ changes from 1 to 0, this means classifying from fine to coarse, integrating in turn, and finally getting the dynamic clustering graph. Then, the corresponding λ value is selected on-demand, and the corresponding R_λ is combined to clarify the requirements category and the corresponding Kano type of personalized requirements. The fuzzy clustering analysis is carried out by MATLAB, and the results are summarized. The details are as follows (Table 5).

Table 5 Results summary of Kano fuzzy classification

Kano type	Initial personalized demand items
M	3
O	1 2 6 7 8 9 11 13 14 17 20
A	4
I	5 10 12 15 16 18 19 21 22 23 24

The selection of requirement items should be highlighted: Category O can be retained as a customizable item for satisfaction reasons as a desired requirement with a high level of attention. Category A is an unexpected demand item that contributes to satisfaction and can therefore be retained as a customizable item. So, all items except categories A and O do not need to be retained. So far, we obtain the bottom requirements of the hierarchical of needs model through the above steps.

4 Importance Ranking of Personalized Demand Items

4.1 Calculating Initial Weights of Personalized Demand Items for Car Sharing

(1) In the screening of personalized needs, 12 items are finally retained. In order to clarify the importance of these projects, it is necessary to prepare a special questionnaire for further investigation. Still using the sampling method used in the previous Kano survey, 100 questionnaires were issued and 83 valid questionnaires were recovered, with an effective rate of 83%.

(2) Based on the survey results, the initial weight w_j of each demand is solved by the entropy method. The steps of the entropy method are as follows:
First: Construct decision matrix $A = (a_{ij})_{m \times n}$. Let m customers evaluate the importance of n personalized demand items. a_{ij} represents the rating of customer i for personalized demand item j, which is used as the basis for the construction of the decision matrix A.
Second: normalize the decision matrix A. Normalize the decision matrix A to $B = (b_{ij})_{m \times n}$ using $b_{ij} = a_{ij} / \sum_{i=1}^{m} a_{ij}$.
Third: Calculate the entropy value. Let the entropy value of personalized demand term j be e_j, then $e_j = \sum_{i=1}^{m} b_{ij} \ln b_{ij}, i = 1, 2, \cdots, m, j = 1, 2, \cdots, n$.
Fourth: Determine the initial weight of the personalized demand item. Let $u_j = \frac{1}{e_j}, j = 1, 2, \cdots, n$. When u_j is normalized, the initial weight of item j is $w_j = u_j / \sum_{j=1}^{n} u_j$, and $\sum_{i=1}^{j} w_j = 1$.
The calculation results of the initial weight w_j are shown in Table 6.

Table 6. Result of initial weight and weight adjustment

Demand items FRi	Side airbags	Vehicle data recorder	Emergency call for accident	Adaptive cruise control	Parking assist	Lane keeping system
wj	0.088	0.082	0.090	0.083	0.087	0.088
w'i	0.081	0.076	0.083	0.154	0.087	0.081
Demand items FRi	Behavior detection system	Air cleaning system	Automotive connectivity ecosystem	Electric sliding door	Smart NFC key	In-car Wi-Fi
wj	0.090	0.083	0.080	0.078	0.078	0.076
w'i	0.082	0.077	0.074	0.072	0.072	0.069

4.2 Weight Adjustment and Determining the Importance Ranking of Personalized Demand items for car Sharing Functions

When considering the importance of each demand, we should not only pay attention to the importance evaluation given by customers, but also take into account the role of satisfaction improvement. Therefore, based on the Kano model, the process of weight adjustment is redesigned to obtain the weight results of each requirement. The specific adjustment methods are as follows:

(1) Find the corresponding adjustment factor k according to the Kano category. A parametric function can be used to represent the correlation between the realization of demand and customer satisfaction in the Kano model. The formula is as follows:

$$s = cx^k \tag{5}$$

In the formula, s is customer satisfaction, c represents a constant, x reflects the level of demand satisfaction, k represents the adjustment coefficient of various Kano. In this paper, previous studies are used as a reference and the values of k corresponding to M, O, A, I, and R are established as 0.5, 1, 2, 0, and -1 respectively.

(2) Adjust the initial weight. Usually, k value is used as a reference to calculate the weight of all requirements. To represent a personalized requirement i with w'_i, there are:

$$w'_i = \frac{w_i k_i}{\sum_{i=1}^{n} w_i k_i} (i = 1, 2, \cdots, n) \tag{6}$$

In the formula, w_i represents the initial weight of the personalized demand item i and k_i is its corresponding adjustment factor. The process of adjusting the weights is shown in Table 6.

The final personalized demand items can reflect their importance to a certain extent. Then, according to the weight size order, sort out the personalized demand items and import them into the model. See Fig. 1 for details.

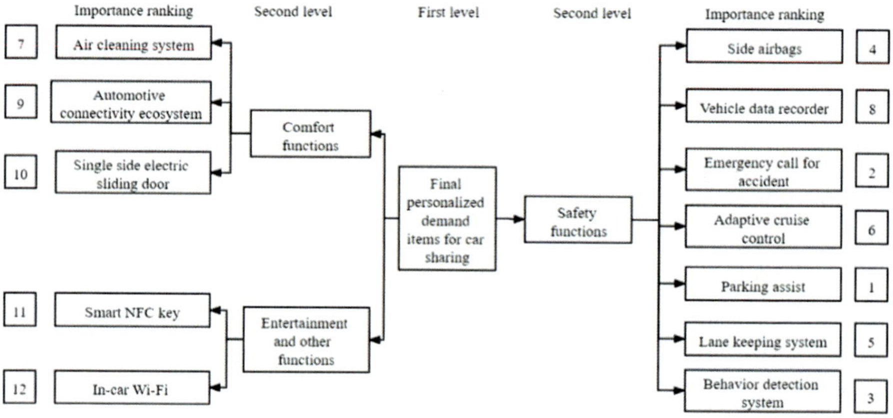

Fig. 1. Final hierarchy of needs model of personalized requirements

5 Conclusion

At present, the demand of car sharing consumers is gradually showing a personalized trend, but the existing car sharing is difficult to meet the needs of a variety of consumers. At the same time, car sharing enterprises can only appropriately select some key personalized needs and give them the opportunity to customize. Therefore, this paper intends to analyze the advantages of car sharing customization, and determine the product customization attributes and customization priorities, to balance the contradiction between customer satisfaction and product production and maintenance costs. Firstly, a personalized demand acquisition method based on Kano model is proposed. Based on the principle of fuzzy theory, the Kano model is introduced to screen the personalized demand items, so that the personalized demand items of car sharing are presented in hierarchy of needs model, and their importance ranking is determined. The research results tell us that the overall importance of safety functions is the highest. This shows that consumers have high demand and expectations for the safety customization function of car sharing. When encountering conflicts of attributes or requirements, the customization of demand items should be arranged step by step based on the order of importance.

Acknowledgements. This work was supported by the National Natural Science Foundation of China (NO. 52072131), the Key Research Projects of Universities in Guangdong Province (NO. 2019KZDXM009), the Science and Technology Project of Guangzhou City (NO. 201804010466), and the Special Innovative Projects of Universities in Guangdong Province (NO. 2019GKTSCX036).

References

1. Ortúzar, J.: Future transportation: Sustainability, complexity and individualization of choices. Commun. Transp. Res. **1**, 100010 (2021). https://doi.org/10.1016/j.commtr.2021.100010
2. Keigo, A., Fuminori, S., Junichiro, O.: Impacts of ride and car-sharing associated with fully autonomous cars on global energy consumptions and carbon dioxide emissions. Technol. Forecast. Soc. Chang. **174**, 121311 (2022)
3. Wang, Y., Wu, J., Chen, K., Liu, P.: Are shared electric scooters energy efficient? Commun. Transp. Res. **1**, 100022 (2021). https://doi.org/10.1016/j.commtr.2021.100022
4. Chaoru, L., Liu, C.: Ecological control strategy for cooperative autonomous vehicle in mixed traffic considering linear stability. J. Intell. Connected Veh. **4**(3), 115–124 (2021). https://doi.org/10.1108/JICV-08-2021-0012
5. Yusuke, K., Hisatomo, H., Takashi, O.: Impacts of shared autonomous vehicles: tradeoff between parking demand reduction and congestion increase. Transp. Res. Interdisc. Perspect. **12**, 100482 (2021)
6. Jiaming, W., Kulcsár, B., Selpi, X.Q.: A modular, adaptive, and autonomous transit system (MAATS): an in-motion transfer strategy and performance evaluation in urban grid transit networks. Transp. Res. Part A Policy Pract. **151**, 81–98 (2021). https://doi.org/10.1016/j.tra.2021.07.005
7. Xu, Y., Ye, Z., Wang, C.: Modeling commercial vehicle drivers' acceptance of advanced driving assistance system (ADAS). J. Intell. Connected Veh. **4**(3), 125–135 (2021). https://doi.org/10.1108/JICV-07-2021-0011
8. Chen, A., Lu, Y.: Protective behavior in ride-sharing through the lens of protection motivation theory and usage situation theory. Int. J. Inf. Manag. **61**, 102402 (2021). https://doi.org/10.1016/j.ijinfomgt.2021.102402
9. Barbara, H., Eva, H.: The social dilemma of car sharing - the impact of power and the role of trust in community car sharing. Int. J. Sustain. Transp. 1–24 (2021). https://doi.org/10.1080/15568318.2021.1912224
10. Konstantin, K., Michael, K., Kay, W.: What drives the utility of shared transport services for urban travellers? A stated preference survey in German cities. Travel Behav. Soc. **26**, 206–220 (2022)
11. Fatma, I., Şeniz, Ç.: Integrating the Kano model into architectural design: quality measurement in mass-housing units. Total Qual. Manag. Bus. Excell. **26**(3–4), 400–414 (2015)

12. Gao, K., Yang, Y., Li, A., Li, J., Yu, B.: Quantifying economic benefits from free-floating bike-sharing systems: a trip-level inference approach and city-scale analysis. Transport. Res. Part A Policy Pract. **144**, 89–103 (2021)
13. Fei, M., Dan, G., Kum, F., Qipeng, S., Fuxia, R., Xiaobo, X., Chengyong, Z.: The influence of continuous improvement of public car-sharing platforms on passenger loyalty: a mediation and moderation analysis. Int. J. Environ. Res. Public Health **17**(8), 1–21 (2020)
14. Jerry, W., Arvind, R.: Customerization: the next revolution in mass customization. J. Interact. Mark. **15**(1), 13–32 (2010)
15. Tseng, M.M., Radke, A.M.: Production planning and control for mass customization – a review of enabling technologies. In: Fogliatto, F.S., da Silveira, G.J.C. (eds.) mass customization, pp. 195–218. Springer, London (2011). https://doi.org/10.1007/978-1-84996-489-0_10
16. Tan, K.C., Shen, X.X.: Integrating Kano's model in the planning matrix of quality function deployment. Total Qual. Manag. **11**(8), 1141–1151 (2000)

Characteristics Extraction and Increasing Block Fine Modeling for Repeated Speeding Behaviors

Yuan Yao[1], Chuanyun Fu[2](✉), Guifu Li[2], and Yajie Li[1]

[1] School of Transportation and Logistics, Southwest Jiaotong University, Chengdu, China
[2] School of Transportation Science and Engineering, Harbin Institute of Technology, Harbin, China
fuchuanyun@hit.edu.cn

Abstract. To efficiently deter the repeated speeders who are frequently fined but continue to commit the violation, this study attempts to investigate the characteristics of the repeated speeding behaviors and propose an increasing block fine modeling approach. Based on the off-site law enforcement data collected from the Deyang City, three speeding ranges (low ($\leq$20%), medium (20%–50%), and high ($\geq$50%)) were considered, and then the characteristics of repeated speeding behaviors were extracted. After that, the cost-benefit theory was introduced to develop an increasing block fine model by taking into account the speeding range and frequency. Considering the minimum total number of speeding as the goal and the economic cost as the constraint, an optimization model of increasing block fine was developed. The pattern search algorithm was used to solve the developed optimization model to determine the best number of blocks and corresponding fine within each speeding range. Finally, a case study was conducted to validate the performance of the developed model. The results show that for the repeated speeders, the percent of the low speeding behavior is the highest, and the speeding behaviors largely occur during the daytime. Furthermore, within the increasing block fine mechanism, the reduction rates of medium and high speeding behaviors are clearly higher than that of the low speeding behavior. The findings of this study offer a fresh viewpoint on the repeated speeding intervention and enable the speeding enforcement more equitable.

Keywords: Repeated speeding behavior · Increasing block fine · Cost-benefit theory · Price elasticity · Traffic violation

1 Introduction

Speeding behavior refers to the driving speed of a vehicle that exceeds the speed specified by law and regulations [1]. In China, traffic accidents caused by speeding are among the top traffic offenses [2]. In other countries, speeding is also responsible for approximately one-third of road fatalities. Related studies focused on the characteristics of speeders (such as gender [3–5], age [6, 7], personality [8–10], behavior [11], and so on) and the consequences of the accident [12]. However, few studies have explored the characteristics of speeding behavior, especially for repeated speeding behavior.

Y. Bie et al. (Eds.): KES-STS 2022, SIST 304, pp. 55–64, 2022.
https://doi.org/10.1007/978-981-19-2813-0_6

Bahram et al. [13] found that drivers with repeated speeding experiences were more likely to speed again. Manderso et al. [14] believed that repeat speeders are more likely to exceed the speed limit by 20 km/h in the process of speeding. Moreover, studies display that serious speeders are more likely to commit speeding again [15, 16].

To reduce speeding, the corresponding penalty mechanism was proposed and implemented. However, the current punishment scheme for speeding only takes into account the radius of action. Namely, the greater the speeding range, the heavier the punishment. It is worth noting that the current penalty system for speeding offenses does not take into account the number of times of speeding, making it difficult to efficiently deter repeated offenders [17]. This leads to a lack of fairness in traffic enforcement.

In order to deal with repeat traffic offenders, academics are learning from the economics scale theory [18]. The stepwise increasing penalty models of illegal parking and red-light running are constructed successively [19–22]. This provides a theoretical reference for the construction of the stepwise increasing penalty model of repeated speeding. However, the step-up penalty model for illegal parking and red light running only takes into account the number of violations. In fact, the stepwise penalty model should not only consider the speeding range [23] but also take into account the number of illegal times under different speeding ranges [24], which needs further studies.

Therefore, based on the off-site law enforcement data, extracting repeated speeding behavior characteristics, introducing the theory of ladder pricing, the increasing block fine model of repeated speeding behaviors with the consideration of both speeding range and frequency is developed, and then a case study is conducted to verify the model performance. The results of this study help not only to deter offenders who have been repeatedly punished for speeding but also to enhance the fairness of traffic law enforcement.

2 Off-Site Law Enforcement Data Collection and Preprocessing

The off-site law enforcement data collected from 2017 from the Traffic police Detachment of Deyang Public Security includes the illegal vehicle license plate, illegal content, vehicle type, occurrence time and location, and other information. Based on illegal content, the keyword "exceeding the prescribed speed limit" was used to extract the data of speeding offenses. After deleting duplicate and invalid data, 179,221 speeding violations were recorded, and a total of 114,976 vehicles were involved. Among them, 110,093 speeding data were labeled with speed limit conditions. Further, the speeding range was divided into: low (less than 20%), medium (greater than or equal to 20% but less than 50%), and high (greater than or equal to 50%).

3 Speeding Behavior Feature Extraction

3.1 Speeding Frequency Within Each Range

Among the 179, 221 speeding data of Deyang, there were 132064 low-range speeding behaviors; 44,206 medium-range speeding behaviors; and 2,951 high-range speeding behaviors. For 100,096 items, the speed limit situation of the speeding location is indicated, and the data is counted according to different speeding ranges and different speed limit values, as shown in Fig. 1. It can be seen that under low-speed limit conditions, drivers are more likely to have high-range speeding behaviors; under high-speed limit conditions, drivers are more likely to have low-range speeding behaviors and less frequent high-range speeding behaviors.

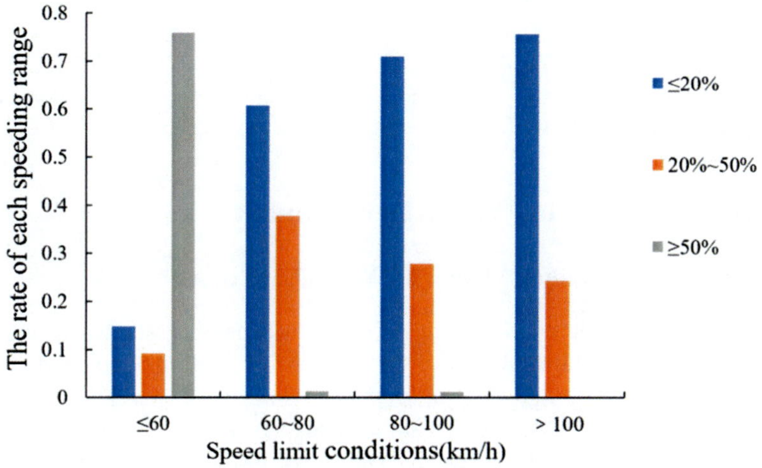

Fig. 1. The rate of each speeding range under different speed limit conditions

3.2 Speeding Range and Interval Time

The distribution of single and repeated speeding behaviors with different speeding ranges is shown in Table 1. It can be seen that the percentages of repeated speeding behaviors with different ranges are greater than those of single speeding behaviors.

Table 1. Distribution of single and repeated speeding behaviors with different speeding ranges.

Speeding range	Single		Repeated		Total
	Number	Percentage	number	Percentage	
Low (≤20%)	58153	44.03%	73911	55.97%	132064
Medium (20~50%)	20604	46.61%	23602	53.39%	44206
High (≥50%)	1473	49.92%	1478	50.08%	2951

As shown in Table 2, among the repeated speeding behaviors, two types of speeding behaviors occurred the most. With the increase of the number of speeding, the proportion of speeding behaviors gradually decreased. Because China currently adopts a point deduction mechanism for speeding behaviors, too many penalty points will make drivers lose their licenses, which has a great inhibitory effect on repeated speeding behaviors.

Table 2. The times of speeding behavior

Violation times	Number of speeding behaviors	Percentage
1	80230	69.78%
2	20810	18.10%
3	7401	6.44%
4	3100	2.70%
5	1483	1.29%
6	774	0.67%

4 Increasing Block Fine Model Development

4.1 The Number of Speeding Violations Within Each Block

Based on the research on the characteristics of speeding behavior, a stepwise incremental penalty model considering the number of steps and the proportion of covered drivers is constructed.

Use P_i to indicate the original speeding range i ($i = 1, 2, 3$, corresponding to low, medium, and high ranges, respectively). After the newly implemented tiered fines mechanism, the amount of change in the fines at all levels of the speeding range is as follows: $\Delta P_i = [\Delta P_{i1}, \Delta P_{i2}, \ldots, \Delta P_{im}]$. Then the newly implemented step fines will be fined at all levels of the speeding range is $P_a = P_i + \Delta P = [P_i + \Delta P_{i1}, P_i + \Delta P_{i2}, \ldots, P_i + \Delta P_{im}]$ (m corresponds to the order). In the formulation of the number of steps, the proportion of the number of speeding drivers is considered. At low stairs, it covers the vast majority of drivers. As the number of steps increases, the number of drivers covered by a single step gradually decreases. The number of steps and the reference value of coverage is shown in Table 3.

Table 3. Reference values for the numbers of steps and coverage rate

Number of steps	Speeding interval of each order			
	First step	Second step	Third step	Fourth step
2	(0, 1]	$(1, +\infty)$	—	—
3	Cover 70% or 80% of speeding drivers	Accumulated coverage of 90% or 95% of speeding drivers	Cumulative coverage 100% Speeding drivers	—
4	(0, 1]	Accumulatively cover 70% or 80% of speeding drivers	Accumulated coverage of 90% or 95% of speeding drivers	Cumulative coverage 100% speeding drivers

4.2 Model Construction of Fines

1) **The elasticity of fines for motorists' demand for speeding**
   In the theory of supply and demand, the price elasticity of demand is used to describe the sensitivity of consumer demand to changes in commodity prices. such as:

$$\frac{\Delta Q}{Q} = E\frac{\Delta P}{P} \tag{1}$$

   where Q is the demand before price adjustment; E is the price elasticity coefficient; ΔQ is the change in demand after price adjustment; ΔP is the price adjustment amount.

   According to the theory of supply and demand, the driver's demand for speeding (speeding times) will change as the amount of fines changes. When the speeding fine is increased, the number of speeding of the driver will decrease accordingly; conversely, the number of speeding will increase.

2) **The response of a single driver to the increasing block fine mechanism for speeding behavior**
   According to formula (1), after adjusting the fine limit, the amount of change in the number of speeds after the driver responds is:

$$\Delta Q_i = E\frac{\Delta P_{ij}}{P_i}Q_i \tag{2}$$

   Where i =[1, 2, 3] is the speeding range; P_i is the fine amount of i under the speeding range before the adjustment; ΔP_{ij} is the speeding range i, the adjustment amount of the j-th order fine.Q_i is the number of excessive speeds of the driver in the speeding range i before the adjustment.

   The number of speeding times when the driver is in the speeding range level i $Q_i > m$. The fine adjustment amount under the speeding range i of this order is ΔP_{im}. After adjustment, the amount of change in the number of speeding times and the number of speeding times after the response of the driver in the speeding range i are as follows:

$$\Delta Q_i = E\frac{\Delta P_{i1}}{P_i}N_1 + E\frac{\Delta P_{i2}}{P_i}(N_2 - N_1) + \cdots + E\frac{\Delta P_{im}}{P_i}(Q_i - N_{m-1}) \tag{3}$$

$$Q_{aim} = Q_i + E\frac{\Delta P_{i1}}{P_i}N_1 + E\frac{\Delta P_{i2}}{P_i}(N_2 - N_1) + \cdots + E\frac{\Delta P_{ij}}{P_i}(Q_i - N_{m-1}) \tag{4}$$

3) **Changes in fines after the driver responds**
   According to the driver's response to the speeding behavior in a stepwise increase of fines, it can be seen that: After the driver responded, the number of speeding times under the speeding range level i changed from Q_i to Q_{ai}, at this time, the speeding rank to which the driver belongs also changes. After responding, the fines faced by the driver of the new m-level speeding times under the speeding range level i are:

$$F_{aim} = (\Delta P_{i1} + P_i)N_1 + (\Delta P_{i2} + P_i)(N_2 - N_1) + \cdots + (\Delta P_{im} + P_i)(Q_{aim} - N_{m-1}) \tag{5}$$

4) **The response of all drivers to the increasing block fine mechanism for speeding behavior**
The response of the driver in the speeding range i to the increasing block fine mechanism for speeding behavior. Before the implementation of the increasing block fine mechanism for speeding behavior, the total number of speeding Q_{si} and the total fine F_{si} of the driver under the speeding range i can be obtained by the relevant points of $f(Q_i)$ in the entire speeding range:

$$Q_{si} = \int_{0}^{+\infty} f(Q_i)Q_i dQ_i \tag{6}$$

$$F_{si} = \int_{0}^{+\infty} f(Q_i)P_i Q_i dQ_i \tag{7}$$

In the model, it can be approximated that drivers with the same number of speeding have the same response to this fine mechanism. After implementation, the total number of speeding Q_{sai} and the total fine F_{sai} of drivers in the speeding range i can be obtained by summing the responses of drivers of all levels:

$$Q_{sai} = \int_{0}^{N_1} f(Q_i)Q_{ai1} dQ_i + \int_{N_1}^{N_2} f(Q_i)Q_{ai2} dQ_i + \cdots + \int_{N_{m-1}}^{+\infty} f(Q_i)Q_{aim} dQ_i \tag{8}$$

$$F_{sai} = \int_{0}^{N_1} f(Q_i)F_{ai1} dQ_i + \int_{N_1}^{N_2} f(Q_i)F_{ai2} dQ_i + \cdots + \int_{N_{m-1}}^{+\infty} f(Q_i)F_{aim} dQ_i \tag{9}$$

5) **A stepwise increasing fine limit optimization model based on driver response**
According to the fine elasticity of the driver's speeding demand, and its response to the implementation of a step-by-step penalty mechanism for speeding behavior, it is possible to establish a stepwise increasing fine limit optimization model for speeding behavior. The goal of the model is to minimize the number of speeding in the study area after implementation, that is, the objective function is:

$$\min Q_{sa} = \sum_{i=1}^{3} w_i \left[\int_{0}^{N_1} f(Q_i)Q_{ai1} dQ_i + \int_{N_1}^{N_2} f(Q_i)Q_{ai2} dQ_i + \cdots + \int_{N_{m-1}}^{+\infty} f(Q_i)Q_{aim} dQ_i \right] \tag{10}$$

When implementing this stepped incremental fine mechanism, the total fines for speeding drivers in each speeding range should be greater than the total fines before the implementation. But it should not exceed the driver's acceptable range. At the same time, under the same speeding range level of the model, the fines of each level should be increased sequentially; taking into account the more serious impact of the high speeding range, the fine for each speeding range should be greater than the fine for the lower speeding range and smaller than the fine for the higher speeding range.

4.3 Optimization of Fines for Each Level

By introducing the calibrated parameters into the fine limit optimization model and considering the number of speeding times, a stepwise increasing fine limit optimization model for repeated speeding behaviors can be proposed as follows:

$$\min Q_{sa} = \sum_{i=1}^{3} w_i[\int_{0}^{+\infty} f(Q_i)Q_i dQ_i + \int_{0}^{+\infty} f(Q_i)E\frac{\Delta P_{i1}}{P_i}Q_i dQ_i +$$

$$\int_{1}^{+\infty} f(Q_i)E\frac{\Delta P_{i2} - \Delta P_{i1}}{P_i}(Q_i - 1)dQ_i + \int_{3}^{+\infty} f(Q_i)E\frac{\Delta P_{i3} - \Delta P_{i2}}{P_i}(Q_i - 3)dQ] \quad (11)$$

Where Eq. (11) represents the goal which is to minimize the number of speeding behaviors in the study area after implementation.

5 Case Analysis

Based on traffic violation data and traffic violation penalties in Deyang City, a case analysis is carried out on the constructed stepwise increasing fine limit optimization model. Using the pattern search method to solve the model in MATLAB, obtain the amount of change in the amount of the fines at all levels of the speeding range compared to the original fines that only consider the speeding range is $\Delta P = [10, 30, 60, 20, 60, 100, 60, 200, 350]$. Then the amount of the fines at all levels corresponding to each speeding range is $P = [110, 130, 160, 170, 210, 250, 560, 700, 850]$. Therefore, the corresponding stepwise incremental penalty mechanism is shown in Fig. 2.

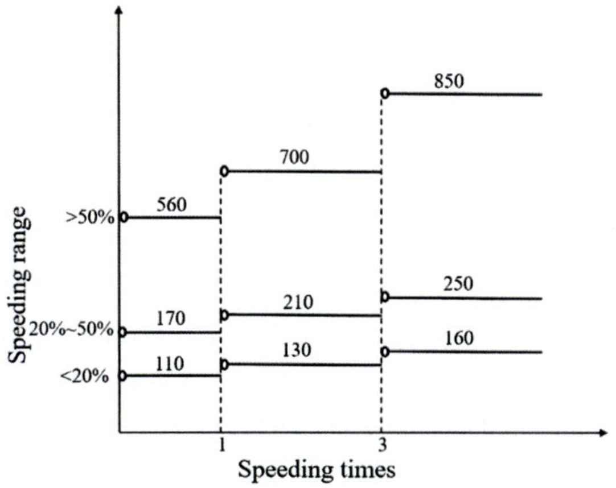

Fig. 2. Increasing block fine mechanism of the speeding behavior.

By comparing the driver's total number of speeding times and the original number of speeding times after the incremental penalty mechanism was tested, the rate of change of the number of speeding times is obtained, as shown in Table 4. It can be seen from Table 4 that in each speeding range, as the number of speeding steps increases, the number of speeding decreases, and the percentage of decrease gradually increases. The rate of decrease in the number of speeding in the middle and high speeding ranges is significantly higher than that in the low speeding range, especially under the second and third levels of speeding.

Table 4. Changes of speeding number in each step due to the incremental fine

Speeding range	Speeding block	Speeding interval	Original speeding times	Speeding times after response	The change rate of speeding times
Low range <20%	1	1	58153	57462	−1.19%
	2	2~3	47292	45604	−3.57%
	3	>3	26619	24337	−8.57%
Mid range20%~50%	1	1	20604	20278	−1.58%
	2	2~3	15488	14736	−4.85%
	3	>3	8114	7232	−10.87%
High range >50%	1	1	1473	1452	−1.42%
	2	2~3	1043	995	−4.65%
	3	>3	435	388	−10.76%
Total	—	—	179221	172485	−3.76%

6 Conclusions and Future Research Directions

The speed limits are found to be significantly associated with speeding behavior; Repeated speeding behaviors account for about 55% of speeding violations; most of repeated speeding drivers will speed again within a month.

A stepwise incremental fine limit and optimization model that concurrently considers speeding frequency and severity is constructed to determine the best step number and fine limit for each step, so that the penalty for speeding violations is fairer.

The case analysis shows that the stepwise incremental penalty mechanism has a significant deterrent effect on repeated speeding behaviors.

In China, as the times of speeding increase, the drivers might become more careful while driving since they are worried about losing their licenses because of accumulated penalty points. This may impact the inhibitory effect of increasing block fine mechanism on repeated speeding behaviors, which should be accounted for in future works.

In the future, the fines for repeated speeding behaviors can be further adjusted by considering the speeding-related outcomes such as crash risk (e.g., potential crashes and traffic conflicts [25–27]). As well, the microscopic traffic simulation approach [28] can be applied to investigate the effectiveness of increasing block fine strategy.

Acknowledgment. This work was supported by the National Natural Science Foundation of China (Grant No. 71801182) and the Fundamental Research Funds for the Central Universities (Grant No. FRFCU5710000111).

References

1. Kong, X., Das, S., Jha, K., Zhang, Y.: Understanding speeding behavior from naturalistic driving data: applying classification based association rule mining. Accid. Anal. Prev. **144**, 105620 (2020)
2. Fu, C., Liu, H.: Investigating distance halo effect of fixed automated speed camera based on taxi GPS trajectory data. J. Traffic Transp. Eng. (English Edition) 1–30 (2021)
3. Fu, C., Li, Y., Hu, Z.: Fine modes of repeat speeding behaviors. China Saf. Sci. J. **29**(03), 14–20 (2019). (in Chinese)
4. Atumo, E.A., Jiang, X., Fu, C.: Spatial point pattern analysis of traffic violations in Luzhou City China. Transp. Lett. **13**(10), 1–10 (2021)
5. Tseng, C.: Operating styles, working time and daily driving distance in relation to a taxi driver's speeding offenses in Taiwan. Accid. Anal. Prev. **52**, 1–8 (2013)
6. World Health Organization: Speed management. https://www.who.int/publications-detail-redirect/managing-speed (2017)
7. Zhang, G., Yau, K.W., Gong, X.: Traffic violations in Guangdong province of China: speeding and drunk driving. Accid. Anal. Prev. **64**, 30–40 (2014)
8. Ismutulla, E., Li, X., Ilzat Eli, I.: Field test-based study on physiological and psychological characteristics of drivers on plateau roads. China Saf. Sci. J. **25**(12), 16–21 (2015). (in Chinese)
9. Elder, R.W., et al.: recommendations on the effectiveness of ignition interlocks for preventing alcohol-impaired driving and alcohol-related crashes. Am. J. Prev. Med. **40**(3), 362–376 (2010)
10. Pan, C., Xu, J., Fu, J.: Effect of gender and personality characteristics on the speed tendency based on advanced driving assistance system (ADAS) evaluation. J. Intell. Connected Veh. **4**(1), 28–37 (2021). https://doi.org/10.1108/JICV-04-2020-0003
11. Zahid, M., Chen, Y., Khan, S., Jamal, A., Ijaz, M., Ahmed, T.: Predicting risky and aggressive driving behavior among taxi drivers: do spatio-temporal attributes matter? Int. J. Environ. Res. Public Health **17**(11), 3937 (2020)
12. Li, Z., Chen, X., Ci, Y., Chen, C., Zhang, G.: A hierarchical Bayesian spatiotemporal random parameters approach for alcohol/drug impaired-driving crash frequency analysis. Anal. Methods Accid. Res. **21**, 44–61 (2019)
13. Goldenbeld, C., Reurings, M., Van Norden, Y., Stipdonk, H.: Crash involvement of motor vehicles in relationship to the number and severity of traffic offenses. An exploratory analysis of Dutch traffic offenses and crash data. Traffic Injury Prev. **14**(6), 584–591 (2013). https://doi.org/10.1080/15389588.2012.743125
14. Qiao, J., Li, W., Guo, R., Li, Z.: Insight into the curved running speed, the driver's gaze and heart-beat temp features of the large trucks on the 2-way 4-lane highways. J. Saf. Environ. **21**(2), 700–706 (2021). (in Chinese)
15. Zhou, Y., Jiang, X., Fu, C., Liu, H.: Operational factor analysis of the aggressive taxi speeders using random parameters Bayesian LASSO modeling approach. Accid. Anal. Prev. **157**, 106183 (2021)

16. Ravani, B., Wang, C.: Speeding in highway work zone: an evaluation of methods of speed control. Accid. Anal. Prev. **113**, 202–212 (2018)
17. Manderson, J.L., Siskind, V., Bain, C., Watson, B.: Speeding recidivism and road safety. In: 2004 Road Safety Research, Policing and Education Conference, pp. 1–8, Road Safety Council of Western Australia, Perth, WA (2004)
18. Watson, B., Watson, A., Siskind, V., Fleiter, J., Soole, D.: Profiling high-range speeding offenders: Investigating criminal history, personal characteristics, traffic offences, and crash history. Accid. Anal. Prev. **74**, 87–96 (2015)
19. Gao, K., Yang, Y., Li, A., Li, J., Yu, B.: Quantifying economic benefits from free-floating bike-sharing systems: a trip-level inference approach and city-scale analysis. Transp. Res. Part A Policy Pract. **144**, 89–103 (2021). https://doi.org/10.1016/j.tra.2020.12.009
20. Zeng, Y.: Research on step punishment intensity model in automated enforcement (in Chinese). Master degree dissertation, Harbin Institute of Technology. (2014)
21. Faruqui, A., George, S.: Quantifying customer response to dynamic pricing. Electr. J. **18**(4), 53–63 (2005)
22. Fu, C., Xiong, Y., Zhang, Y., Zhang, W., Liu, Y.: A novel model of increasing block fine for red-light running recidivism. Adv. Mech. Eng. **11**(2), 1–13 (2019)
23. Xu, Y., Ye, Z., Wang, C.: Modeling commercial vehicle drivers' acceptance of advanced driving assistance system (ADAS). J. Intell. Connected Veh. **4**(3), 125–135 (2021). https://doi.org/10.1108/JICV-07-2021-0011
24. Liu, H.: Research on optimization model of the increasing tiered fine for motorized vehicle running red-light. Master degree dissertation, Harbin Institute of Technology (2016). (in Chinese)
25. Fu, C., Tarek, S.: Bayesian dynamic extreme value modeling for conflict-based real-time safety analysis. Anal. Methods Accid. Res. **34**, 100204 (2022)
26. Fu, C., Tarek, S., Zheng, L.: Multi-type Bayesian hierarchical modeling of traffic conflict extremes for crash estimation. Accid. Anal. Prev. **160**, 106309 (2021)
27. Fu, C., Tarek, S.: Multivariate Bayesian hierarchical Gaussian copula modeling of the non-stationary traffic conflict extremes for crash estimation. Anal. Methods Accid. Res. **29**, 100154 (2021)
28. Gressai, M., Varga, B., Tettamanti, T., Varga, I.: Investigating the impacts of urban speed limit reduction through microscopic traffic simulation. Commun. Transp. Res. **1**, 100018 (2021). https://doi.org/10.1016/j.commtr.2021.100018

An Effective Berths-Based Approach to Calculate the Capacity of Drop-Off Exclusive Roadway

Yaping Zhang[1], Guifu Li[1], Chuanyun Fu[1](✉), and Qian Luo[2]

[1] School of Transportation Science and Engineering, Harbin Institute of Technology, Harbin 150090, China
fuchuanyun@hit.edu.cn
[2] Chengdu Civil Aviation Information Technology Co., Ltd., Chengdu 610041, China

Abstract. To accurately estimate the capacity of multi-lane drop-off exclusive roadway, a calculation approach based on effective berths is proposed. First, the study specifically analyzes different layout types of parking and through lanes and introduces the concept of effective berths of drop-off exclusive roadway. A calculation approach of the capacity of multi-lane drop-off exclusive roadway at airport terminal based on effective berth and spatiotemporal trajectory theory is proposed. The effective berth under a certain level of delay is determined via VISSIM simulation and the drop-off exclusive roadway in Jinan Yaoqiang Airport-China is taken as an example for case analysis. The results show that the accuracy of the proposed method is more than 95%. The proposed approach is more in line with the actual unload process of the drop-off vehicles than traditional methods and lays a theoretical foundation for service level evaluation and the management and guidance of drop-off vehicles.

Keywords: Traffic capacity · Effective berth · Drop-off exclusive roadway · Spatiotemporal trajectory theory · Airport terminal

1 Introduction

In recent years, due to the rapid growth of the number of passengers traveling by air and the limitation of airport land space, many cities are extending or constructing airports. In this context, how to rationally design the layout of landside facilities to improve the corresponding traffic capacity is one of the main problems to be resolved. As one of the airport's roadside facilities, the drop-off exclusive roadway refers to the road set up in front of the departure hall of the terminal building for vehicles to complete passage and short stops for passengers to disembark and unload luggage. There are flows of passengers and vehicles which intertwine on the staging area (i.e., the section outside the terminal). Especially for the peak period, the congestion and traffic disorder are widespread and seriously affect traffic efficiency of the airport's landside (i.e., the drop-off exclusive roadway). How to accurately estimate the traffic capacity of a drop-off

Y. Bie et al. (Eds.): KES-STS 2022, SIST 304, pp. 65–74, 2022.
https://doi.org/10.1007/978-981-19-2813-0_7

exclusive roadway is a key issue for accurately evaluating its service level, effectively managing drop-off vehicles, which requires in-depth exploration.

At present, scholars have carried out a lot of studies on the calculation approach of the capacity of the drop-off exclusive roadway at the airport terminal, mainly focusing on theoretical and simulation methods. De Neufville and Odoni [1] defined the drop-off exclusive roadway and gave the empirical formula for calculating the length of the drop-off roadway in "Airport System: planning, design, and management". Zhang et al. [2] analyzed the characteristics of traffic flow on the drop-off exclusive roadway and proposed calculation methods of traffic indicators (including average vehicle stay time, average number of vehicles on the drop-off exclusive roadway, etc.) based on the queuing theory. Chen et al. [3] proposed a lane-changing behavior based on a deep learning approach to simulate the relationship between the speed and spacing of the lane-changing vehicle. Liu et al. [4] developed a model to control the demand scale of the drop-off exclusive roadway based on the multi-desk queuing theory with the capacity limitation (M/M/n/C/). Parizi and Braaksma [5] established a basic mathematical calculation model for the capacity of the drop-off exclusive roadway with a single parking lane, which ignores the deceleration behavior of vehicles parking and driving away. Zhang et al. [6–8] proposed formulas for calculating the capacity of two-lane drop-off exclusive roadway by considering the change of speed and the interaction between internal and external lanes. Zhang [9] analyzed the capacity of a three-lane drop-off exclusive roadway with double parking lanes. The introduction of connected and autonomous vehicles will reshape the transport system. Traffic flow, particularly the road capacity and free-flow speed could considerably be improved [10, 11]. Carrone et al. [12] investigated how the utilization of the road capacity degraded as a function of heterogeneity in congested motorways.

Javid et al. [13] proposed the conceptual model of the simulation analysis of the drop-off exclusive roadway, laid a theoretical foundation for the subsequent simulation system practice. Chang [14] developed a multi-lane parking simulation model with driver preferences in mind. ACRP and TRB [15] gave a method to estimate the required length of parking lane in "Airport Cooperative Research Program Report 40 (ACRP40)" and suggested VISSIM simulation could be used to estimate the capacity of drop-off exclusive roadway.

The above-mentioned works mostly focus on the two-lane drop-off exclusive roadway with one parking lane. Currently, the drop-off exclusive roadways lanes of most terminals are more than three. For multi-lane drop-off exclusive roadways with different layouts of lane functions, each has its operation regulation, and the utilization rate of parking lanes is also diverse. There are varying degrees of mutual interference between vehicles. The capacity of the drop-off exclusive roadway is not equal to the mere accumulation of the capacity of every single lane. Therefore, the existing capacity calculation methods are not applicable anymore.

This study starts from the differences in the layout of parking and through lanes of drop-off exclusive roadway at airport terminal, summarizes the lane function layout types, and proposes a calculation approach of the capacity of multi-lane drop-off exclusive roadway based on effective berth and spatiotemporal trajectory theory. Effective berth at each delay level under a certain level of delay is determined via VISSIM simulation, and a case study is conducted to verify the accuracy of the capacity calculation approach proposed in this study. The proposed approach is suitable for various numbers of lanes and lane function layout types and provides a new method for calculating the capacity of drop-off exclusive roadways.

2 Lane Function Layout Types of Drop-Off Exclusive Roadway

Drop-off exclusive roadways are composed of single or multi-lane groups. Each lane group includes one parking lane and one through lane at least. Drop-off vehicles make a short stop and drop off on the parking lane, and complete the process of entering and leaving the drop-off area on the through lane. Figure 1 is a schematic diagram of a double-width drop-off exclusive roadway.

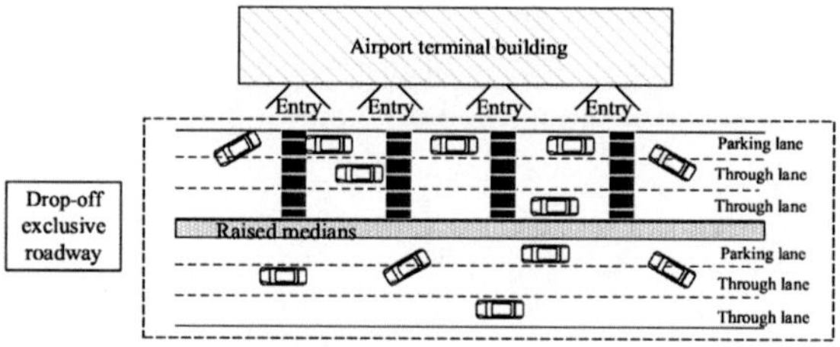

Fig. 1. The layout of the curbside area alone the drop-off exclusive roadway.

The drop-off exclusive roadway can be divided into double-lane, triple-lane, four-lane, and five-lane according to the number of lanes. Two and three-lane groups are used primarily at small or medium airports, and four and five-lane groups are used at large airports. The layout types of drop-off exclusive roadways are also diverse, which are summarized into 9 types, as shown in Table 1.

Table 1. Lane function layout types of drop-off exclusive roadway.

Number of lanes	Layout type		
Double-lane	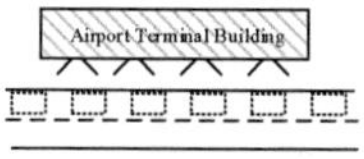 1P1T		
Triple-lane	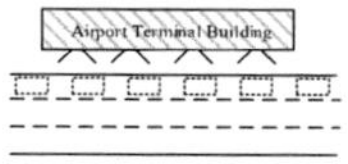 1P2T	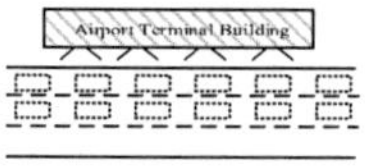 2P1T	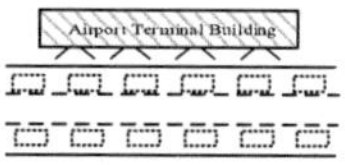 2P1T*
Four-lane	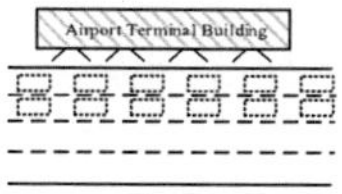 2P2T	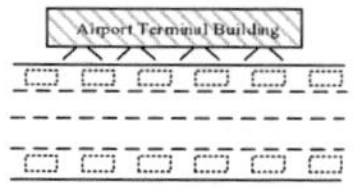 2P2T*	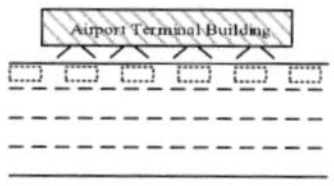 1P3T
Five-lane	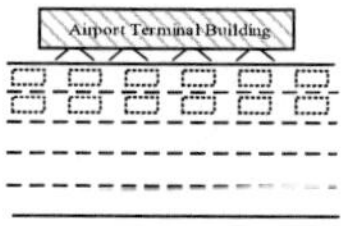 2P3T	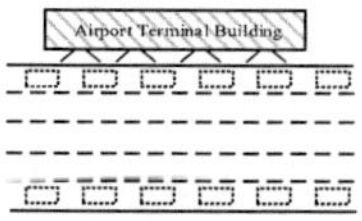 2P3T*	

Note: P stands for the parking lane; T represents the through lane; * represents the middle lane.

3 Calculation Approach of the Capacity of Multi-lane Drop-Off Exclusive Roadway Based on Effective Berths

3.1 The Theoretical Model Based on Time-Space Trajectory Theory

Parizi and Braaksma [5] have proposed a method for calculating the capacity of drop-off exclusive roadway based on time-space trajectory theory, which mainly considers the following assumptions:

1) Only for two-lane drop-off exclusive roadways.
2) Ignore drop-off vehicles' acceleration and deceleration behaviors.
3) Assuming that the driver has no preference when choosing a parking space.

As shown in Fig. 2, this method divides drop-off exclusive roadway into n linearly arranged parking spaces, and the capacity of drop-off exclusive roadway is the sum of the number of vehicles parked and departed on n parking spaces within a unit time.

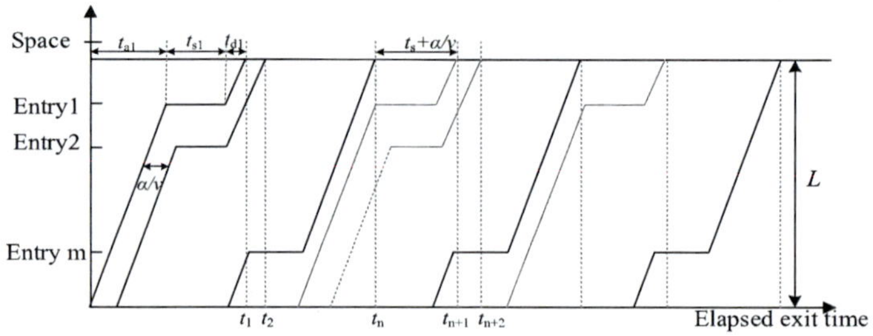

Fig. 2. Time-space diagram of vehicles in case with n entries [5].

Zhang [6] distinguishes the speed of the through lane and the parking lane and takes into account that the vehicle will travel a certain distance on the parking lane and get a modified method, which is constructed as:

$$C = \frac{n(T - \frac{L - L_R}{v_L} - \frac{L_R}{v_R})}{t_s + n\frac{\alpha}{v_R}} \tag{1}$$

$$L_R = \frac{1.007L_E - 2.959}{2} \tag{2}$$

The capacity of the drop-off exclusive roadway is represented by C, where L is the length of the drop-off exclusive roadway; L_R is the driving distance of the vehicle in the parking lane; L_E is the farthest distance between parked vehicles and the entry; v_L represents the speed of the vehicle in the through lane; v_R represents the speed of the vehicle in parking lane; n is the number of parking spaces; T is time, usually, 3600 s; α denotes the average safe distance of vehicles driving in drop-off exclusive roadway; t_s is the time of vehicles parking in the parking lane.

The C obtained by Eq. (1) is the capacity of a two-lane drop-off exclusive roadway (1P1T) in ideal conditions. This calculation method lays the foundation for the calculation approach of the multi-lane drop-off exclusive roadway proposed in this study.

3.2 Concept of Effective Berths

Federal Transit Administration (FTA) [16] first introduced the concept of effective berth and use it to explore the difference between multiple on-line and off-line bus stops. Since then, effective berths have been used to study the transfer capacity of bus stops and the optimal design of bus stops [17]. This study introduces effective berths into the calculation of the capacity of the drop-off exclusive roadway at the airport terminal. The main reasons are as follows:

1) The lane function layout types of drop-off exclusive roadway are diverse, and the number of parking lanes and through lanes are different. The capacity of multiple lanes is not equivalent to the simple accumulation of the capacity of a single lane.
2) Because of the driver's preference, the further the parking lane is, the fewer vehicles are parked and the greater the gaps between parked vehicles are.
3) Vehicles that expect to park on the inner parking lane often need to cross the outer parking lane from the through lane to park, which often occupies the outer parking lane's time and space resources available for parking.
4) When a vehicle parking inner prepares to leave, the distance and enough time for the vehicle to pass need to be considered. Sometimes it is necessary to wait for the outer vehicle to leave before looking for a gap to drive away.

Therefore, the effective berth factor of drop-off exclusive roadway is defined as: under a certain delay level, the ratio of the number of vehicles served by the multi-lane drop-off exclusive roadway and the number of vehicles served by the standard drop-off exclusive roadway (1P1T) within a unit time, which can be seen in Eq. (3):

$$N_{ei} = Q_i / Q_1 \tag{3}$$

The effective berth factor of the drop-off exclusive roadway is represented by N_{ei}, where Q_1 is the number of vehicles served by the standard drop-off exclusive roadway with 1 parking lane and 1 through lane within a unit time; Q_i is the number of vehicles served by the multi-lane drop-off exclusive roadway within a unit time.

3.3 Calculation Approach of the Capacity of Drop-Off Exclusive Roadway

The formula for calculating the capacity based on effective berth is:

$$C_i = N_{ei} \frac{n\left(T - \frac{L-L_R}{v_L} - \frac{L_R}{v_R}\right)}{t_s + \frac{n\alpha}{v_R}} \tag{4}$$

For multi-carriageway drop-off exclusive roadway, the capacity is the sum of the capacity of the lane groups.

4 Determination of the Effective Berth Factors Based on Simulation

This section simulates 9 lane function layout types of the drop-off exclusive roadway via VISSIM. To analyze each type, the vehicle input of each type and the corresponding delay in the simulation are recorded, and then draw scatter diagrams of the simulation output data. Use Eq. (3) to calculate efficient berth factors for different types of lane arrangements on exclusive drop-off roadway under a certain level of delay.

4.1 Basic Assumptions and Parameter Settings

When calculating effective berth factors, in addition to limiting a certain level of delay, it is also necessary to ensure that the traffic of drop-off exclusive roadways is comparable. This simulation ensures that the main boundary conditions of all drop-off exclusive roadways are the same, that is, vehicles have the same operating characteristics. The VISSIM simulation is based on the following assumptions:

1) Vehicles on exclusive roadways are all cars of the same size and parameters.
2) The drop-off vehicle's arrival obeys the same negative exponential distribution.
3) The parking time of vehicles on obeys the same distribution.
4) The length and width of each lane are identical.
5) Ignore the pedestrian and vehicle interference caused by the pedestrian crossing.

The simulation parameters used in the simulation are listed in Table 2:

Table 2. Parameters used in the simulation.

Simulation parameter	Value
Expected speed of vehicles on through lanes	30 km/h
Expected speed of vehicles on parking lanes	5.5 km/h
Parking time distribution	Empirical distribution (6~225 s)
Vehicle type	100% cars
Length of drop-off exclusive roadway	270 m
Lane width	3.5 m
The standard number of parking spaces of a parking lane	40
Length of per parking spaces	6 m
Priority rules	Vehicles on the through lane first

4.2 Simulation Results Analysis

The simulation software VISSIM is used to establish the operation simulation systems to analyze the drop-off exclusive roadways. Delay measurements are set in 9 sub-models, and the attributes of results are output. Delay is equal to the actual travel time on the drop-off exclusive roadway minus the theoretical (ideal) travel time.

Set the initial value of the traffic flow on the drop-off exclusive roadway to start the simulation, and record the vehicle inputs and delays for each simulation. In the first stage, the vehicle input is added by 100 pcu/h in turn, and the average delay of the vehicle is recorded after each simulation. After that, add the vehicle input until the simulation gives a capacity limit warning that “the vehicle cannot be input completely”. On the basis of the input of vehicles before the warning appears, add the vehicle input 10 pcu/h in turn, until the capacity limit warning appears again. Then, the vehicle input is added by 1 pcu/h in turn until the warning appears again, which means the traffic volume of the drop-off exclusive roadway in the simulation system has reached its peak and the drop-off exclusive roadway has reached its capacity.

From the simulation results, we found that the capacity of drop-off exclusive roadway was not directly proportional to the number of parking lanes, but was related to the number of through lanes and the layout of lane functions. Select the two-lane drop-off exclusive roadway (1P1T) as the base type, analyze the simulation results of the three-lane, four-lane, and five-lane drop-off exclusive roadway respectively. Take the average delay as the x-axis, and the vehicle arrival rate (vehicle input) as the y-axis, then draw the data scatter diagram of each lane group, as shown in Fig. 3.

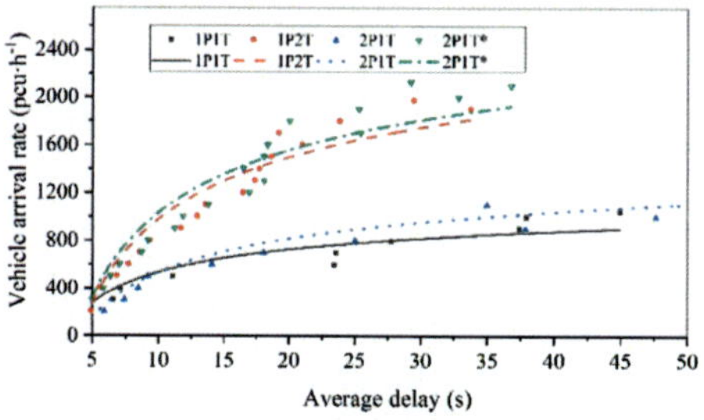

(a) three-lane drop-off exclusive roadway

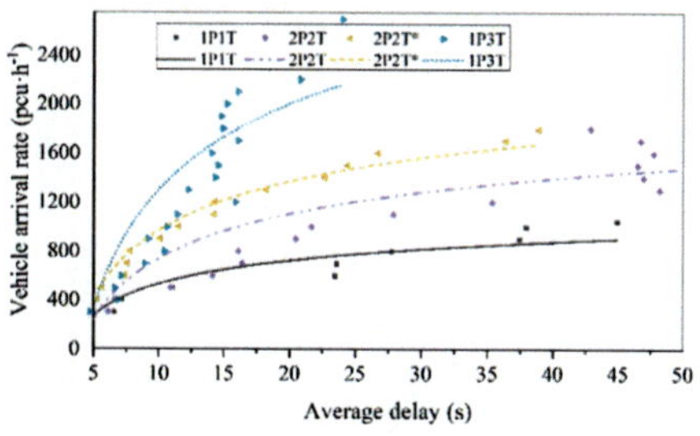

(b) four-lane drop-off exclusive roadway

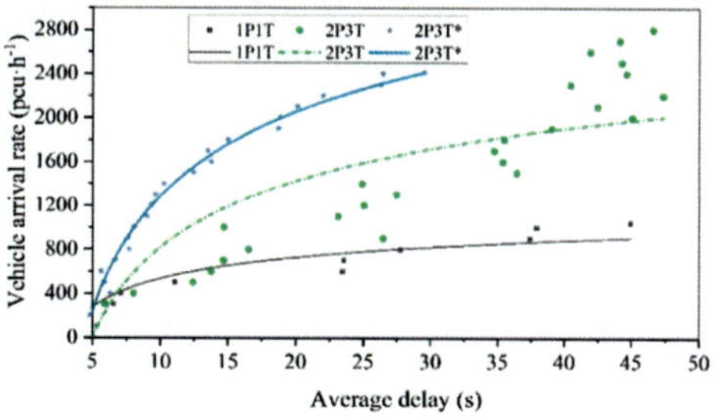

(c) five-lane drop-off exclusive roadway

Fig. 3. Comparative analysis of simulation results of each lane group.

Drop-off exclusive roadways have an upper limit of their capacity. According to the model of vehicle arrival rate and delay given by Zhang et al. [18], combined with the changing trend of the curve in Fig. 3, it can be concluded that when the vehicle arrival rate reaches the upper limit of the load-bearing capacity of the drop-off exclusive roadway, the average delay is about 25 s. The arrival rate-delay curves intersect at some points on the delay interval (5, 10), and use the new methodology proposed by Wang et al. [19]

to calibrate. Use Eq. (3) to calculate the effective berth factors for various lane groups when the average delay is 25 s, as shown in Table 3.

Table 3. Effective berth factors of each type.

IDX	Qty of lane	Lane function layout	Qty of P	Qty of T	Effective berth factor
1	Double-lane	1P1T	1	1	1.00
2	Triple-lane	1P2T	1	2	2.08
3		2P1T	2	1	1.14
4		2P1T*	2	1	2.17
5	Four-lane	2P2T	1	3	1.53
6		2P2T*	2	2	1.88
7		1P3T	2	2	2.76
8	Five-lane	2P3T	2	3	2.01
9		2P3T*	2	3	2.88

5 Case Analysis

To verify the accuracy of the calculation approach of the capacity of multi-lane drop-off exclusive roadway, this study takes Jinan Yaoqiang International Airport-China as a case to analyze the capacity of drop-off exclusive roadway.

The drop-off exclusive roadway at Jinan Yaoqiang Airport adopts the type of lane function layout of 1 parking lane and 2 through lanes arranged in sequence (1P2T), the effective berth factor found in Table 4 is 2.08. Use Eq. (4) to calculate the capacity of the drop-off exclusive roadway at Jinan Yaoqiang Airport. The values of specific parameters are shown in Table 4, which are determined based on survey results.

Table 4. Values of parameters in calculating drop-off exclusive roadway capacity.

Parameter	L/m	N_{ei}	n	t_s/s	L_R/m	$v_L\ /km{\cdot}h^{-1}$	$v_R\ /km{\cdot}h^{-1}$	α/m
Value	270	2.08	37	61.2	30	30	10	6

Substitute the above parameters into Eq. (4), the capacity of the inner drop-off exclusive roadway is 1942 pcu/h. According to the actual survey results, due to the driver's preference, the ratio of traffic volume between the inner and outer roadways is about 2:1, so the capacity of the drop-off exclusive roadway is as follows:

$$C = C_{inner} + C_{outer} = (1 + 0.5)C_{inner} = 1.5 \times 1942 = 2913\,\text{pcu/h} \tag{5}$$

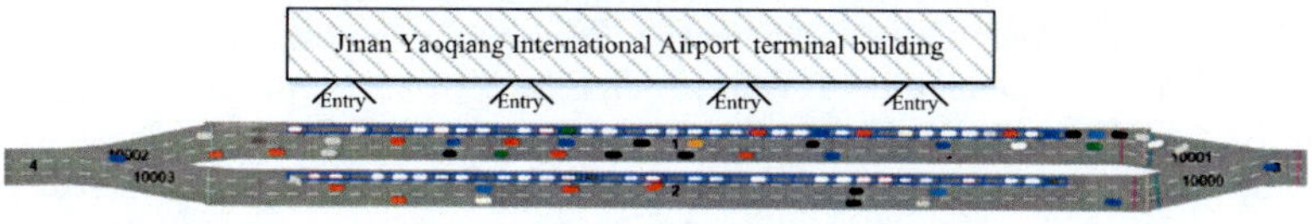

Fig. 4. VISSIM simulation model of Jinan Yaoqiang Airport drop-off exclusive roadway.

Then simulates this drop-off exclusive roadway via VISSIM, as shown in Fig. 4. When the capacity limit warning appears, the maximum input is 3049 pcu/h. Compared with the calculated capacity (2913 pcu/h), the relative error is 4.3%. The results verify the accuracy of the effective berth-based calculation approach of the capacity of the drop-off exclusive roadway at the airport terminal.

6 Conclusions

In this study, the capacity of drop-off exclusive roadway at airport terminal is studied, and an approach for calculating the capacity for multiple lane numbers and different lane function layout types is proposed. The lane function layout types of drop-off exclusive roadway are sorted out, the concept of effective berth is introduced, and the capacity calculation approach of multi-lane drop-off exclusive roadway based on effective berth is proposed. VISSIM simulation is used to determine the effective berth factors under different delay levels, and a case study at the terminal of Jinan Yaoqiang Airport is carried out. The results show that the accuracy of the calculation approach for the capacity based on effective berth is as high as 96%. Compared to traditional methods, the proposed approach for calculating capacity is more consistent with the parking characteristics of vehicles on drop-off exclusive roadways.

However, the drop-off area in front of the airport terminal is very complex. The proposed methodology ignores the influence of pedestrian and vehicle interference and the arrival distribution, which may affect the accuracy of the capacity calculation. Therefore, the mutual interference, conflicts between pedestrians and vehicles [20–22], and the arrival distribution [23] should be considered in future studies. Furthermore, drop-off exclusive roadway performance under mixed connected and regular vehicle environments is also a meaningful direction to explore [24].

Acknowledgment. The work was jointly supported by the National Natural Science Foundation of China (Grant No. 71801182), the Sichuan Science and Technology Program (Grant No. 2019JDTD0001), and the Fundamental Research Funds for the Central Universities (Grant No. FRFCU5710000111).

References

1. de Neufville, R., Odoni, A.: Airport system: Planning, design and management (Aviation Week Book). Library of congress Cataloging-in-Publication, Washington (2003)
2. Zhang, L., Wang, Z., Fang, S.: Study on traffic capacity and demand of airport terminal curbside. J. Tongji Univ. (Nat. Sci.) **35**(04), 486–490 (2007). (in Chinese)

3. Liu, W., Zhou, H.: Analysis to traffic capacity of departure curbside parking of passenger transportation in airport landside. J. Transp. Sci. Eng. **26**(02), 98–102 (2010). (in Chinese)
4. Chen, K., Liu, P., Li, Z., Wang, Y., Lu, Y.: Modeling anticipation and relaxation of lane changing behavior using deep learning. Transp. Res. Rec. **2675**(12), 186–200 (2021)
5. Parizi, M.S., Braaksma, J.P.: Optimum design of airport enplaning curbside areas. J. Transp. Eng. **120**(4), 536–551 (1994)
6. Zhang, Z.: Capacity evaluation of hub airport landside curbside. Master degree dissertation, Civil Aviation University of China (2019). (in Chinese)
7. Yang, F., Li, T.: Traffic characteristics and capacities of passenger drop-off area at large intermodal transportation. J. Zhejiang Univ. (Eng. Sci.) **51**(11), 2207–2214 (2017). (in Chinese)
8. Lu, X., Zhu, J., Tang, X.: Capacity assessment and optimization of terminal curbside. J. Harbin Inst. Technol. **41**(09), 96–99+135 (2009). (in Chinese)
9. Zhang, Y.: Analysis on dynamic capacity of curbside on transportation hub. Technol. Econ. Areas Commun. **15**(04), 78–81 (2013). (in Chinese)
10. Liu, Y., Guo, J., Taplin, J., Wang, Y.: Characteristic analysis of mixed traffic flow of regular and autonomous vehicles using cellular automata. J. Adv. Transp. **2017**, 8142074 (2017)
11. Talebpour, A., Mahmassani, H.S.: Influence of connected and autonomous vehicles on traffic flow stability and throughput. Transp. Res. Part C Emerg. Technol. **71**, 143–163 (2016)
12. Carrone, A.P., Rich, J., Vandet, C.A., An, K.: Autonomous vehicles in mixed motorway traffic: capacity utilisation, impact and policy implications. Transportation **48**(6), 2907–2938 (2021)
13. Javid, M., Bergener, J.M., Seneviratne, P.: Airport curbside traffic management: a conceptual model for evaluating alternative strategies. In: 75th TRB Annual Meeting, Washington (1994)
14. Chang, K.: A simulation Model for Analyzing Airport Terminal Roadway Traffic and Curbside Parking. Bachelor degree dissertation, University of Maryland (2001)
15. National Academies of Sciences, Engineering, and Medicine, Transportation Research Board, Airport Cooperative Research Program: Airport curbside and terminal area roadway operations. The National Academies Press, Washington (2010)
16. National Academies of Sciences, Engineering, and Medicine, Transportation Research Board: Bus rapid transit, volume 2: Implementation guidelines. National Academies Press, Washington (2004)
17. Gong, X., Wei, Z.: Analysis of effective number of berths at bus stops based on VISSIM. J. Transp. Inf. Saf. **27**(02), 140–142 (2009). (in Chinese)
18. Zhang, L., Jiao, P., Si, M.: Delay model of vehicles at urban road drop-off area. J. Syst. Simul. **32**(09), 1839–1846 (2020). (in Chinese)
19. Wang, S., Chen, X., Qu, X.: Model on empirically calibrating stochastic traffic flow fundamental diagram. Commun. Transp. Res. **1**, 100015 (2021)
20. Fu, C., Sayed, T.: Bayesian dynamic extreme value modeling for conflict-based real-time safety analysis. Anal. Methods Accid. Res. **34**, 100204 (2021)
21. Fu, C., Sayed, T., Zheng, L.: Multi-type Bayesian hierarchical modeling of traffic conflict extremes for crash estimation. Accid. Anal. Prev. **160**, 106309 (2021)
22. Liu, Y., Alsaleh, R., Sayed, T.: Modeling lateral interactions between motorized vehicles and non-motorized vehicles in mixed traffic using accelerated failure duration model. Transp A: Transp. Sci. 1–24 (2021)
23. Wu, J., Liu, P., Zhou, Y., Yu, H.: Stationary condition based performance analysis of the contraflow left-turn lane design considering the influence of the upstream intersection. Transp. Res. Part C Emerg. Technol. **122**, 102919 (2021)
24. Zhang, Z., Yang, X.: Analysis of highway performance under mixed connected and regular vehicle environment. J. Intell. Connected Veh. **4**(2), 68–79 (2021)

Impact Analysis of Wired Charging and Wireless Charging on Electric Bus Operation: A Simulation-Based Method

Wei Qin, Libing Liu, Jinhua Ji, Mingjie Hao, and Yiming Bie(✉)

School of Transportation, Jilin University, Changchun 130022, Jinlin, China
yimingbie@126.com

Abstract. In recent years, many cities in the world are committed to promoting the electrification of public transportation. For bus companies, how to select the right charging facilities accurately and quickly has become an urgent problem to be solved. In this paper, we propose a simulation method based on Anylogic to describe the operation of electric buses under wired charging and wireless charging conditions. We provide decision-making suggestions for bus companies by analyzing the impact of wireless charging and wired charging on operation cost and passenger waiting time. According to the simulation results, we found that the waiting time of passengers under wired charging conditions is about 8.63% higher than that under wireless charging in the same operating conditions. The use of wireless charging facilities can effectively reduce the waiting time of passengers.

Keywords: Charging facilities · Electric buses · Passengers waiting time · Simulation

1 Introduction

Electric buses (EBs) have the advantages of zero emission and low noise, which is of great significance to reduce urban motor vehicle exhaust emission and the operation cost of public transport enterprises [1]. A recent study by Bloomberg new energy finance electric predicts that the number of electric buses in operation will double from 386000 in 2017 to 1.2 million, accounting for more than 47% of the total number of urban buses in the world by 2025 [2]. Although EBs have many advantages and develop rapidly, they still have limitations such as short driving range and long charging time [3]. In order to maintain the normal operation of EBs and improve their operational efficiency, the optimization of EBs charging facilities has become an urgent problem for public transport companies.

To solve the charging problem of EBs, there are three charging technologies are used at present: station-based charging [4], battery swapping [5] and wireless lane-based charging [6]. The charging facilities are divided into two categories: wired charging facilities and wireless charging facilities. Wired charging facilities are the most common at present, which have good stability and controllability. However, due to complex

Y. Bie et al. (Eds.): KES-STS 2022, SIST 304, pp. 75–84, 2022.
https://doi.org/10.1007/978-981-19-2813-0_8

equipment operation and slow charging speed, wired charging facilities can only be set up in the terminal stations or depots, which limits the charging accessibility of EBs and requires EBs to equip large capacity batteries [7]. The wireless charging facility adopts WPT technology. EBs can be charged without cables and connectors. The wireless charging facility is convenient, and does not have spark and electric shock risk, in addition, the charging speed of it is very fast. They can be set not only at the terminal bus stations, but also at intermediate stations; By setting up multiple wireless charging facilities in bus stations, the battery capacity of EBs can be reduced and the charging efficiency of EBs will be greatly improved [8].

For the selection and optimization of charging facilities, current studies are mainly based on the theoretical analysis model and verified by practical operation [9, 10]. However, due to the shortage of funds, time, materials and so on, it is difficult for researchers to conduct tests directly in the early research stage, which makes it impossible to evaluate the impact of different facilities on bus operation efficiency and passenger satisfaction. Compared with the real experiment, the virtual experiment save a lot of cost and time. In previous studies, Hao et al. developed a dynamic programming model that optimally schedules the bus operating speed at road sections and multiple signal timing plans at intersections to improve bus schedule adherence [11]. Shi et al. used the pre-established Anylogic urban dynamics model to simulate the hourly power demand of private electric vehicles considering population, commerce, housing and transportation infrastructure, and solved the problem of power imbalance [12]. Although many articles used virtual experiments to solve problems, no simulation experiment is used to study the selection of charging facilities.

Anylogic is a simulation modeling software that supports agent-based modeling. The software has specific industry libraries such as process library, pedestrian library and road traffic library, which can meet the needs of EBs simulation experiment. Therefore, based on Anylogic simulation software, this paper constructs the operation status of EBs under wired charging and wireless charging respectively; The program is written with the built-in module of Anylogic to analyze the impact of different charging facilities on passenger waiting time and charging cost, so as to provide suggestion in the selection of charging facilities of EBs.

The structure of the rest of this paper is as follows: the second part expounds the basic operation strategy, the third part analyzes the example, and the fourth part is the conclusion.

2 Problem Description

2.1 Problem Environment Descriptions

It is assumed that only one EBs line is operating on the public transit exclusive lane. The upward direction of the line is represented by u and the downward direction is represented by d; There are $2N$ stations along the line, n is the station number, $n = 1, 2..., N$. The number of initial station and terminal station is 1 and N respectively. It is assumed that the maximum number of EBs which can be put into operation is K, k is the number of EBs, $k = 1, 2, \ldots, K$; the rated battery capacity of EB k is B_k (unit: kWh), the remaining battery capacity of EB k is B'_k, and the average energy consumption per hundred kilometers of EB k is C_k (unit: kWh/km).

If EBs use wired charging facilities, a wired charging station will be established at station 1 (i.e. the initial station). Assume that there are M wired charging piles in the starting station, and m is the serial number of wired charging piles, $m = 1, 2, \ldots, M$. It is assumed that the 0–1 variable P_c^m to judge whether the wired charging pile is in use. If the wired charging pile m is used, $P_c^m = 1$; otherwise, $P_c^m = 0$. The charging power of the wired charging pile m is P_w^m (unit: kW).

If EBs use wireless charging facilities, the wireless charging pad will be built at the midway stations. Assuming that the 0–1 variable $F_{n,u}$ is the use state of the wireless charging pad at the upward line, if the wireless charging pad is used at the upward station n, set $F_{n,u} = 1$; otherwise, set $F_{n,u} = 0$. The power of the wireless charging pad at upward station n is $F_w^{n,u}$ (unit: kW). The expression of downward stations is in a same way.

If the driver's rest time T_d is satisfied, EBs operate as far as possible when the available operation time is less than the shift time, the EBs stop operation. Set the total operation shift of each EBs on the same day as X.

2.2 Basic Operation Strategies

Charging Strategy. In the case of laying wired charging facilities, the wired charging pile is only set at the bus departure station. In order to minimize EBs queuing and maintain the health of the battery, EB will charge when the power decreases to a certain level.The minimum state of charge (SOC) for EB to maintain battery health is SOC_y, and the minimum SOC to meet the operation requirements is SOC_l. In order to reduce the charging time, the EB ends the charging behavior when the highest SOC_h is reached each time. It is necessary to judge the charging behavior according to the following strategies every time it returns to the departure station.

When EB arrives at the charging station, enter Step 1.

Step 1: if SOC less than SOC_y, enter Step 2; otherwise, enter Step 6, and set T_{p1} and T_{p2} to 0.
Step 2: if SOC is less than SOC_l and can not be satisfied with the next trip, enter Step 3; otherwise, enter Step 3.
Step 3: if there is no free position for charging pile, otherwise, enter Step 6, and set T_{p1} and T_{p2} to 0.
Step 4: if there is an idle charging pile, EB starts charging and record the charging start time t_{p1} and charging duration T_{p1}, enter Step 5; otherwise, waiting in the charging waiting area and recording the charging waiting time T_{p2}, and then repeat Step 4.
Step 5: EB is charged. If SOC of EB reaches SOC_h, end the charging, record the charging end time t_{p2}, and enter Step 6; otherwise, enter Step 4.
Step 6: if $T_{p1} + T_{p2} \geq T_d$, EB starts the next trip; Otherwise, the EBs will leave when the waiting area rests to T_d (Fig. 1).

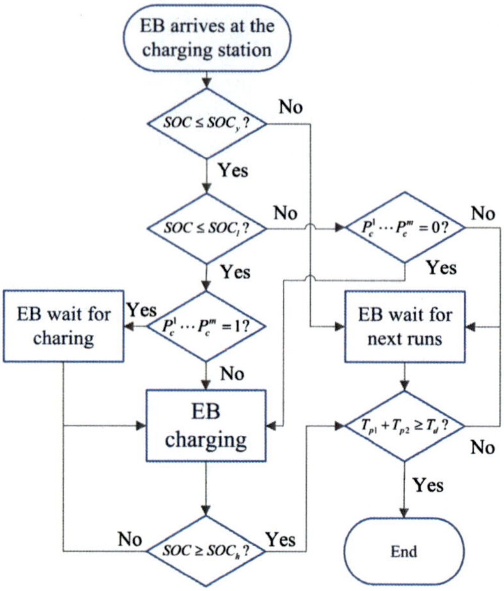

Fig. 1. Wired charging strategy logical flow chart.

In the case of laying wireless charging facilities, the starting station does not need to build charging piles, and the wireless charging pad is set up at the midway station. In order to maintain battery health, EB only uses the time for boarding and alighting when SOC is lower than SOC_y. When EB enters the midway station, the charging strategy will be judged according to the actual situation of passengers getting on and off the bus (Fig. 2):

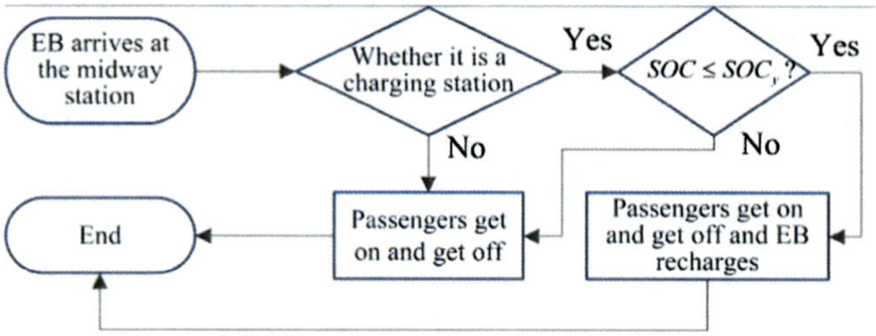

Fig. 2. Wireless charging strategy logical flow chart.

Step 1: when EB arrives at the bus midway station, judge whether the station is equipped with a wireless charging pad. If so, enter Step 2; otherwise, EB will load and unload passengers normally.
Step 2: if SOC of EB is less than SOC_y, enter Step 3; otherwise, EB will load and unload passengers normally.

Step 3: EB stops steadily, passengers start getting on and off, wireless charging facilities start working at the same time, and record the start time t_{f1}; the wireless charging behavior stops while the bus service passengers complete and t_{f2} is recorded.

Passenger Simulation Strategy. The passenger simulation strategy is divided into two parts: one is to determine the passenger arrival probability distribution, and the other is to calculate the passenger boarding and alighting time. When the traffic flow density is small and the overall flow is small, the discrete distribution model is more suitable to describe the pedestrian arrival law [13–15], in which Poisson distribution is commonly used to describe the pedestrian arrival law at the bus station.

$$P(s) = \frac{h^s e^{-h}}{s!} \tag{1}$$

where, h is the average number of arriving passengers per minute and $P(s)$ is the probability that there are exactly s passengers per minute.

Passenger boarding and alighting time T_s can be divided into vehicle opening time T_{s1}, passenger boarding and alighting time T_{pa} and bus closing time T_{s2}:

$$T_s = T_{s1} + T_{pa} + T_{s2} \tag{2}$$

$$T_{pa} = \max\{s_u \cdot T_u, s_d \cdot T_d\} \tag{3}$$

where, s_u and s_d are the number of people getting on and getting off, T_u and T_d are the time for each passenger to get on and get off.

Charging Cost Calculation Strategy. Assuming that the electricity price of the city where the line located is time-of-day tariff, there are q electricity price periods, and the cost per kilowatt hour is M_q in period T_q.

$$C = \begin{cases} M_q \cdot T_k^m, & t_k^{m,s} \in \left[t_q^s, t_q^e\right) \text{ and } t_k^{m,e} \in \left[t_q^s, t_q^e\right) \\ \left(t_q^s - t_k^{m,s}\right)M_q + \sum\limits_{\Gamma=q+1}^{q+\varphi-1} (t_\Gamma^e - t_\Gamma^s)M_\Gamma + \left(t_k^{m,e} - t_{q+\varphi}^s\right)M_{q+\varphi}, & t_k^{m,,s} \in \left[t_q^s, t_q^e\right) \text{ and } t_k^{m,e} \in \left[t_{q+\varphi}^s, t_{q+\varphi}^e\right) \end{cases} \tag{4}$$

Where, C is charging cost, T_k^m is charging duration, $t_k^{m,s}$ and $t_k^{m,e}$ are charging start time and end time, t_q^s and t_q^e are charging start time and end time of q electricity price periods, φ is the total electricity price periods.

3 Example Analysis

3.1 Simulation Parameter Setting

In this paper, a bus line in operation is selected for simulation. The line has five midway stations, with a total mileage of 9.5 km and travel time is about 35 min. The line has eight EBs with the same model and the same battery. The battery capacity is 100 kWh and the average energy consumption per 100 km is 80 kWh. Among them, 4 EBs start

Table 1. Distance between stations and average travel time

Section name	s0–s1	s1–s2	s2–s3	s3–s4	s4–s5	s5–s6
Length of road section (m)	1800	1100	1300	1800	1500	2000
Average travel time (min)	2.8	2.6	4.2	5.0	4.2	4.2

*s0 represents departure station and s6 represents terminal. s1, s2, s3, s4 and s5 respectively represent midway station 1, 2, 3, 4 and 5.

from the departure station and the other 4 EBs start from the terminal station. Midway stations are straight-line midway stations. The distance between stations and average travel time are shown in the Table 1.

In this simulation, the number of wired charging pile m is 3, and the charging power of charging pile P_w is 100 kW; The wireless charging pad is set in the upward and downward of station 2, upward of station 4 and downward of station 5 with large passenger flow. The power of the wireless charging pad is 200 kW; Set the driver's rest time to 300 s; Set the maintenance battery health SOC_y to 50%, the SOC_l to 20%, and the SOC_h to 80%.

For the convenience of the study, it is assumed that no passengers get on and off at the departure station, only passengers get on at the midway station, and all passengers get off at the terminal. Set the door opening time T_{s1} as 1.5 s and the vehicle door closing time T_{s2} as 1.5 s; The boarding time T_u of each passenger is 1.8 s and the alighting time T_d is 1.2 s. The passenger flow at the midway station conforms to the Poisson distribution. The average number of passengers arriving at the midway station per minute is shown in the Table 2.

Table 2. Average passengers arrival rate (pas/min)

Direction	1	2	3	4	5
Upward	1.71	2.31	2.00	2.14	1.57
Downward	1.22	2.45	1.76	1.64	1.88

The city uses time-of-day tariff to adjust the electricity price. The specific electricity price and time period are shown in Table 3.

Table 3. Time-of-day tariff table (CNY/kWh)

Time period	7:00–9:00	9:00–11:30	11:30–14:30	14:30–17:00	17:00–19:00
Electricity price	0.975	1.250	1.500	1.250	1.500

3.2 Establishment of Simulation Environment

The simulation steps are divided into simulation scene construction, simulation module connection and data analysis based on Anylogic's own data statistics function. The simulation scene is mainly set up based on Anylogic's process library, pedestrian library and road traffic library. The specific scene setting is shown in the Fig. 3.:

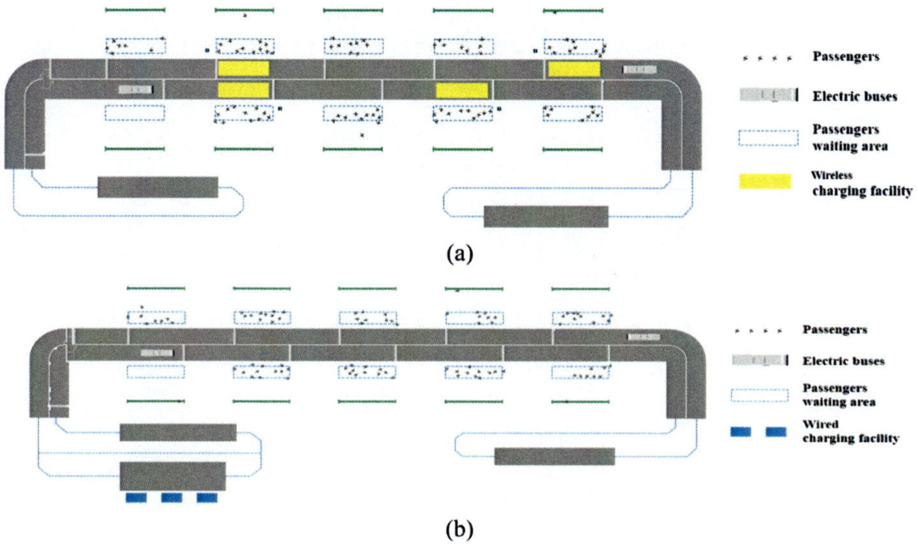

Fig. 3. Laying scenario of wireless charging facilities (a) and wired charging facilities (b)

The simulation module connection is mainly based on Anylogic's process library and pedestrian library. The blue box adopts the process library and the green box adopts the pedestrian library. The specific module connection settings are shown in the Fig. 5.

EBs are produced by the Electric buses source module, moved to the station through the Electric buses move module, and then the Electric buses delay module is used to control the boarding and alighting time of passengers and the Electric buses pick is used to complete the action of obtaining passengers. Passengers are generated by the Passengers source module. Passengers are controlled to move to the station through the Passengers move module, and Passengers wait for boarding by using the Passengers wait module. Finally, the boarding behavior is completed by using the Passengers queue module and Passengers exit module (Fig. 4).

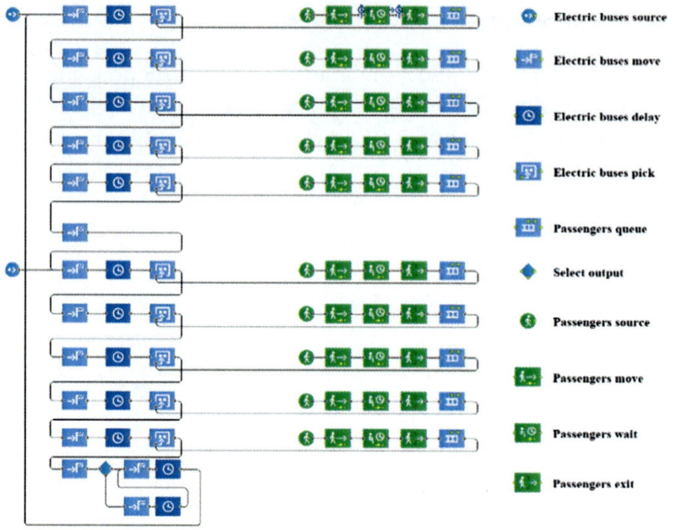

Fig. 4. Connection diagram of simulation module

3.3 Results and Analysis

The model runs from 7:00 a.m. to 19:00 p.m. for a total of 12 h. The waiting time of up and down passengers at each station is analyzed respectively. The results are shown in the Table 4.

Table 4. Waiting time of passengers (s)

Charging facilities	Direction	1	2	3	4	5
Wired charging	Upward	380.4	310.4	390.5	402.3	506.9
	Downward	510.3	530.6	410.3	346.2	339.3
Wireless charging	Upward	367.8	275.1	388.1	368.2	480.6
	Downward	502.4	512.4	379.2	345.2	340.2

It can be seen from the Table 4. that the waiting time of passengers is greater than that of wireless charging under the condition of wired charging. After statistical analysis of all data, the waiting time of passengers under wired charging is about 8.63% higher than that under wireless charging. The use of wireless charging facilities can effectively reduce the waiting time of passengers.

In addition, the total trip of EBs will be different under the influence of different charging methods. Affected by the total trip, the total charging cost of EBs is different. Through data acquisition, the total trip and total charging cost of each vehicle can be obtained, as shown in the Table 5.

Table 5. Total running trips and total charging cost

Vehicle number	Wired charging		Wireless charging	
	Total trip	Total charging cost (CNY)	Total trip	Total charging cost (CNY)
1	13	42.25	14	98.75
2	12	40.75	14	96.88
3	12	76.52	13	90.15
4	12	75.89	13	89.65
5	13	40.25	14	99.25
6	12	43.98	14	99.30
7	12	71.12	13	91.14
8	12	64.28	13	90.45

After statistical analysis, the average total trip of wireless charging is higher 1.25 times than that of wired charging. EB does not need to spend additional time on charging, which effectively makes use of the time for passengers to get on and off, and improves the operation intensity of EBs under wireless charging.

At the same time, the total charging cost of wireless charging is much higher than that of wired charging, and the average charging cost has increased by 37.56. There are two reasons for the increase of charging cost. First, the fleet of wireless charging and wired charging total trip is 108 and 96 respectively. Wireless charging requires more power, resulting in an increase in charging costs. Second, wireless charging is more frequently charged in the peak period of electricity price, while wired charging often occurs in the flat peak period of electricity price, which increases the Total charging cost.

In order to consider the influence caused by the power difference between wireless charging and wired charging, we changed the power of wireless charging facilities to 100 kW and 80 kW. Through simulation, it is found that EB cannot complete daily operation tasks through wireless charging alone. In this case, wired and wireless co-charging is required, so the merits and demerits of wired charging and wireless charging cannot be compared.

4 Conclusions

The construction of EBs charging facilities has always been one of the important directions of EBs research. This paper discusses EBs operation and passenger waiting time under the conditions of wired charging and wireless charging. A bus line is simulated and compared by using Anylogic, which provides a new preliminary investigation method for the construction of bus charging facilities.

This paper only discusses the influence of charging facility selection on EBs from the perspective of simulation, which is still different from the actual operation. However, if you need to further choose charging facilities, it is necessary to conduct more in-depth research in combination with the cost and specific road environment.

Acknowledgements. This study was supported by the National Natural Science Foundation of China (No. 71771062), China Postdoctoral Science Foundation (No. 2019M661214 & 2020T130240), and Fundamental Research Funds for the Central Universities (No. 2020-JCXK-40).

References

1. Pelletier, S., Jabali, O., Mendoza, J, E., et al.: The electric bus fleet transition problem. Transp. Res. Part C Emerg. Technol. **109**, 174–193 (2019)
2. Electric Buses Will Take Over Half the World Fleet by 2025. https://www.bloomberg.com/news/articles/2018-02-01/electric-buses-will-take-over-half-the-world-by-2025. Accessed 01 Feb 2018
3. Jiang, N., Xie, C.: Computing and analyzing mixed equilibrium network flows with gasoline and electric vehicles. Comput. Aided Civil Inf. Eng. **29**(8), 626–641 (2014)
4. Li, J.-Q.: Battery-electric transit bus developments and operations: a review. Int. J. Sustain. Transp. **10**(3), 157–169 (2016)
5. An, K.: Battery electric bus infrastructure planning under demand uncertainty. Transp. Res. Part C Emerg. Technol. **111**, 572–587 (2020)
6. Chen, Z., Yin, Y., Song, Z.: A cost-competitiveness analysis of charging infrastructure for electric bus operations. Transp. Res. Part C Emerg. Technol. **93**, 351–366 (2018)
7. Bi, Z., Keoleian, G.A., Ersal, T.: Wireless charger deployment for an electric bus network: a multi-objective life cycle optimization. Appl. Energy **225**, 1090–1101 (2018)
8. Xu, Y., Zheng, Y., Yang, Y.: On the movement simulations of electric vehicles: a behavioral model-based approach. Appl. Energy **283**, 116356 (2021)
9. Bi, Z., Song, L., De Kleine, R., Mi, C.C., Keoleian, G.A.: Plug-in vs. wireless charging: Life cycle energy and greenhouse gas emissions for an electric bus system. Appl. Energy **146**, 11–19 (2015)
10. Zhang, L., Zeng, Z., Gao, K.: A bi-level optimization framework for charging station design problem considering heterogeneous charging modes. J. Intell. Connected Veh. **5**(1), 8–16 (2022). https://doi.org/10.1108/JICV-07-2021-0009
11. Hao, M., Bie, Y., Zhang, L., Mao, C.: Improving schedule adherence based on dynamic signal control and speed guidance in connected bus system. J. Intell. Connected Veh. (2020). https://doi.org/10.1108/JICV-06-2020-0005
12. Shi, R., Zheng, S., Zhang, C., et al.: Study on EV charging station location planning based on the load balance principle with agent-based AnyLogic simulation. In: The 26th Chinese Control and Decision Conference (2014 CCDC), pp. 1515–1519. IEEE (2014)
13. NadiaS. A., Ghoneim. S. C., Wirasinghe.: Near-side of far-side bus stop: a transit point of view. Transp. Res. Record. Nation Res. Council **761**, 69–75 (1980)
14. Peled, I., Lee, K., Jiang, Y., Dauwels, J., Pereira, F.C.: On the quality requirements of demand prediction for dynamic public transport. Commun. Transp. Res. **1**, 100008 (2021). https://doi.org/10.1016/j.commtr.2021.100008
15. Zhang, W., Zhao, H., Xu, M.: Optimal operating strategy of short turning lines for the battery electric bus system. Commun. Transp. Res. **1**, 100023 (2021). https://doi.org/10.1016/j.commtr.2021.100023

A Data-Driven Method for Diagnosing ATS Architecture by Anomaly Detection

Aimin Zhou[1], Shaowu Cheng[1](✉), Xiantong Li[1], Kui Li[1], Linlin You[2], and Ming Cai[2]

[1] School of Transportation Science and Engineering, Harbin Institute of Technology, Harbin 150090, China
csw_h@hit.edu.cn

[2] School of Intelligent Systems Engineering, Sun Yat-Sen University, Guangzhou 510006, China

Abstract. Autonomous Transport System (ATS) architectures enable a wide range of new applications and bring significant benefits to transport systems. However, during the design stage, errors of the architecture can have an impact on the smooth implementation of the ATS, which will endanger the normal operation of the transport systems. To ensure a high autonomy of the ATS architecture, i.e., "functionally evolvable, logically reconfigurable and physically configurable", the detection of ATS architecture design errors is essential. This paper aims to fill the research gap in the existing research on diagnosing or evaluating ATS architectures. Inspired by word embedding models in natural language processing communities, we propose a data-driven approach to diagnose ATS architectures without prior knowledge or rules. We use an architecture embedding model to generate vector representations of ATS architectures, then train the model through negative sampling of the training dataset to identify the features of abnormal ATS architecture. Finally, we employ the trained model to classify structural errors of the test dataset generated from the ATS architecture. The experimental results show that the proposed method gains a relatively good effect of classifying with an average accuracy of 79.3%, demonstrating the effectiveness of the method.

Keywords: Autonomous Transport System · Architecture embedding model · Triple classification · Vector computation

1 Introduction

With the development of technologies such as self-driving cars and cooperative vehicles infrastructure system, existing transport systems are evolving from intelligent transport systems to autonomous transport systems. To reduce human intervention, autonomous transport systems transport passengers and goods through self-organized operations and autonomous services. All these systems have one common character: Inside the system, independent components have to communicate with others to avoid incomplete or unclear communication mechanisms and inconsistent information quality that complicates the introduction of new services and the involvement of new stakeholders and even blocks this process, it is essential to introduce a unified ATS architecture. An ATS architecture

Y. Bie et al. (Eds.): KES-STS 2022, SIST 304, pp. 85–93, 2022.
https://doi.org/10.1007/978-981-19-2813-0_9

integrates transport services, communications networks, vehicles, transport infrastructure, with traffic participants to provide a steady, trustworthy, secure, and privacy-friendly environment for users. The design process of an ATS architecture involves a great deal of repetitive and specialized work that requires a wealth of knowledge and careful reasoning ability, and analytical skills. The components and interactions of the ATS architecture are in continuous evolution and iteration. The above process may generate new and potential errors in the ATS architecture. Therefore, a scientific diagnosis approach is required to ensure the validity and reliability of the ATS architecture throughout the whole process of architecture evolution and iteration. By diagnosing the ATS architecture designed for a given city or region, potential errors in the architecture can be detected and fixed before implementing the ATS. It is good practice to implement such assurance procedures in the development of designing or modifying an ATS architecture. The diagnosis characteristics can be stored in the computer and reusable. However, no feasible methods that can be applied to the diagnosis process without prior knowledge or rules. In this paper, we propose a data-driven approach that represents the structural features of the ATS architecture through vector representation generated by the ATS architecture embedding model. Then the potential errors in the ATS architecture will be detected by vector computation without additional rules. The method simplifies the anomaly detection process into vector computation, which is relatively suitable for the continuously updated ATS architecture.

The paper is structured as follows: Sect. 2 describes the ATS architecture diagnosis problem and provides an overview of the proposed method. Then our diagnosis approach is presented in Sect. 3. In Sect. 4, we give numerical examples based on dataset from the national ITS reference architecture to demonstrate the effectiveness of the proposed method, and finally, Sect. 5 summarizes this paper.

2 Problem Description and Methodology Overview

Autonomous Transport System transports equipment, traffic participants, goods, information, or resources from one point to another with minimal human intervention [1, 2]. ATS exists in a variety of transportation modes such as trucks, buses, rail, ships, and even aircraft. At the early stage they are typically deployed in controlled industrial areas but are expected to be deployed soon in public areas with various degrees of autonomy. Unlike Autonomous Vehicles (AVs) [3], which offer excellent services to individual passengers, ATS integrates vehicles, freight, traffic participants, infrastructure, and information into a system to meet particular needs. When it comes to ATS architecture, the focus is on the interactions between components in a complex system to ensure the efficiency and the stable operation of ATS.

The ATS architecture is a static framework for giving macro guidance for the transport system [4]. Designing an ATS architecture involves lots of repeatable and technical work that requires fund of knowledge and careful reasoning ability and analytical skills. With continuous evolution and iteration of the ATS architecture, the components and interactions can change accordingly. While unforeknown errors may be generated in the ATS architecture. Detecting potential errors at the architecture design stage can help to reduce cost and improve productivity [5]. The architecture diagnosis checks over the internal logical relations, providing a basis for tracing and solving problems of the ATS architecture. It is vital in ensuring the valid deployment of an ATS through the diagnosis techniques.

With the rise of artificial intelligent technology, diagnosis methods that focus on sensors monitoring and signal processing as core tasks have gradually transitioned to diagnosis approaches based on knowledge [6–8]. The diagnosis knowledge can be stored in the computer and reusable. Therefore, the theoretical ATS architecture will relatively match up with the knowledge-based diagnosis approaches.

Currently, knowledge-based diagnosis methods [9–12] can be broadly classified into two categories: symbolic logic inference methods and representation learning methods. Traditional symbolic logic approaches focus on deterministic deductive reasoning and achieve inference diagnosis by defining ontological axioms or logical rules, which have the advantage of being precise and interpretable. While the main disadvantage is that they require the manual definition of logically strict inference rules, so their coverage is narrow and not easily extended. Besides, they cannot handle implicit or uncertain knowledge. To avoid the manual definition of rules, another kind of symbolic inference method uses statistical models of inductive reasoning to automatically learn rules from a large amount of facts. It can generalize abstract logical rules by learning the common features of tagged cases. The main advantage is that it reduces the workload of manually defining rules, but rule learning consumes too much computing resource and cannot represent implicit or uncertain knowledge either.

The representation learning diagnosis method transforms both entities and relations into the vector space and completes inferential diagnosis by vector computation [13], using low-dimensional dense vectors to represent entities and relations. The parameterized vector is an approximate representation in the vector space based on the existing knowledge in the knowledge graph as a sub-supervised signal. On the other hand, vector-based or neural network based inference computing is an approximate inference result obtained through differentiable representation learning [14]. Therefore, the inference process and result are also uncertain. Thus, representation learning methods and neural network approaches are easier to represent uncertain knowledge and implement uncertain reasoning than logical inference and symbolic rule-based approaches. In addition, the inference is more efficient as the inference process is simplified to a vector computation, eliminating the need for symbolic matching and rule search.

The ATS architecture is a flexible and future-proof architecture that guides the evolution and iteration of the transport system. Therefore, ATS architecture diagnosis also accompanies the whole process of architecture evolution and iteration. The above process may cause unpredictable and artificial errors in the ATS architecture. These changes in architecture caused by the evolution and iteration require updating diagnosis knowledge frequently to accommodate them. Since the dynamic problem is not well solved by traditional logical reference methods, we need a scientific diagnosis approach to ensure validity and reliability. Inspired by the word embedding model in the natural language processing community [15], this paper proposes a data-driven approach to diagnose potential errors without extra knowledge or rules in ATS architectures through our architecture embedding model and vector computation. The method simplifies the inference process into vector computation, overcoming the shortcomings of traditional diagnosis methods that rely on expert knowledge and rules. The method consists of two key components: architecture embedding model and anomaly detection.

3 Architecture Embedding Model

An ATS architecture consists of requisite structural knowledge and constraint rules that provide a guide to establish a relationship with infrastructure, vehicles, traffic participants and system-level functional objects in a transport system. This knowledge can be represented through classical knowledge graph methods, stored in the form of triples as (head entity, relationship, tail entity). Here a single example (roadside devices, provisioning, road network health status detection) means that roadside devices are responsible for provisioning road network health status. While triples are powerful in representing structured data, the symbolic characteristic of such triples makes knowledge graph difficult to handle, especially on a large scale. Due to the above reason and ATS architecture embedding model is proposed, inspired by word embedding models from the natural language processing (NLP) community. The key idea is to vectorize structural details into the continuous vector space, simplifying the difficulty and workload while maintaining the inherent structure of the knowledge graph. The vector representation embedded contains semantic information that can be used in rich downstream applications of NLP such as link prediction and triple classification. The embedding model utilizes rich semantic information about entities and relations, which can significantly improve knowledge acquisition and reasoning ability. The vector representation makes it possible to check whether the triple is correct by vector computation. The criterion of correct triples is defined by the truth value, which is calculated as the Euclidean distance. Architecture anomalies are then identified based on the truth value of the triples. The rationality of the architecture is measured by computing the mean value of all triples in the same vector space. This section consists of two parts: knowledge representation and model training.

3.1 Knowledge Representation in ATS Architecture

The embedding model generates the vector representation of the architectural content based on the co-occurrence distribution of the architectural content in the training dataset. Inspired by the word2vec model proposed by Google in 2013 [13], the architecture embedding model in this paper also uses negative sampling to increase the training speed of the model and improve the quality of the vector representation. Negative sampling training means that the model will be trained on both positive and negative data features during the training process. The architecture embedding model is a three-layer neural network, as shown in Fig. 1, with input, output and hidden layers. It is trained on both positive samples (true connections) and negative samples (incorrect functional and physical object connections). The weights of the hidden layer are a vector representation of a word. To achieve the best effect of the embedding model, this paper defines a loss function to which increases the differentiation between the positive and negative samples by minimization of loss. The trained neural network shows that words that appear in similar contexts will have similar vector representations. In the architecture embedding model, this means that entities co-occurring under the same semantic conditions have similar vector representations.

We obtain an ATS architecture $K = \{(\boldsymbol{e}_i, \boldsymbol{r}_k, \boldsymbol{e}_j)\}$ containing triples, each containing two entities $\boldsymbol{e}_i, \boldsymbol{e}_j \in E$ and the relation $\boldsymbol{r}_k \in R$ between them, where E is the set of entities and R is the set of relations, respectively. Modeling triples we use the TransE model [16], based on its simplicity and effectiveness in achieving state-of-the-art predictive performance. Given a triple instance $(\boldsymbol{e}_i, \boldsymbol{r}_k, \boldsymbol{e}_j)$, the relation $\boldsymbol{r}_k$ in the modeled triple instance is used as a vector transformation from $\boldsymbol{e}_i$ to $\boldsymbol{e}_j$, i.e. when the triple holds, we want to have the effect that $\boldsymbol{e}_i + \boldsymbol{r}_k \approx \boldsymbol{e}_j$. We then score each triple based on $\|\boldsymbol{e}_i + \boldsymbol{r}_k - \boldsymbol{e}_j\|_1$, defining the true value of the triple as:

$$P(\boldsymbol{e}_i, \boldsymbol{r}_k, \boldsymbol{e}_j) = 1 - \frac{1}{\sqrt{3}d}\|\boldsymbol{e}_i + \boldsymbol{r}_k - \boldsymbol{e}_j\|_1 \tag{1}$$

Where d is the dimension of the embedding space. It is clear to see that $P(e_i, r_k, e_j) \in [0, 1]$ because of $\|e_i\|_2 \leq 1$, $\|e_j\|_2 \leq 1$, $\|r_k\|_2 \leq 1$.

Note that $0 \leqslant \|\boldsymbol{e}_i + \boldsymbol{r}_k - \boldsymbol{e}_j\|_1 \leqslant \|\boldsymbol{e}_i\|_2 + \|\boldsymbol{r}_k\|_2 + \|\boldsymbol{e}_j\|_2 \leqslant 3\sqrt{d}$, where the last inequality holds because $\|\boldsymbol{x}\|_1 = \Sigma_i |x_i| \leq \sqrt{d\,\Sigma_i x_i^2} = \sqrt{d}\|\boldsymbol{x}\|_2$, according to the Cauchy-Schwarz inequality.

3.2 Training Model with Negative Sampling

The architectural embedding model is trained with positive and negative samples. We first generate a training set containing all positive samples, and then we make the model work best by defining a minimizing global loss function.

$$\begin{gathered} \min_{\{e\},\{r\}} \textstyle\sum_{f^+ \in \mathcal{F}} \sum_{f^- \in \mathcal{N}_{f^+}} \left[\gamma - P(f^+) + P(f^-)\right]_+, \\ s.t. \|e\|_2 \leq 1, \forall e \in E; \ \|r\|_2 \leq 1, \forall r \in R. \end{gathered} \tag{2}$$

Here $f^+ \in \mathcal{F}$ is a positive sample, $f^- \in \mathcal{N}_{f+}$ is a negative sample constructed from a positive sample, and γ is the boundary condition for determining positive and negative samples. For a triple $(\boldsymbol{e}_i, \boldsymbol{r}_k, \boldsymbol{e}_j)$, we achieve minimization by replacing $\boldsymbol{e}_i$ or $\boldsymbol{e}_j$ randomly with an arbitrary entity $\boldsymbol{e}$ in the entity set, which ensures the triple after the replacement does not exist in the original training set. We believe that the vast majority of triple instances generated in this way are negative samples, using small-batch stochastic gradient descent. The effect of model training is shown in Fig. 2.

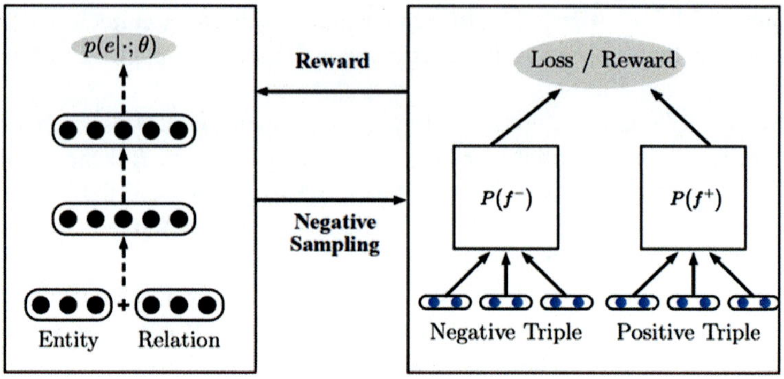

Fig. 1. General structure of the ATS embedding model

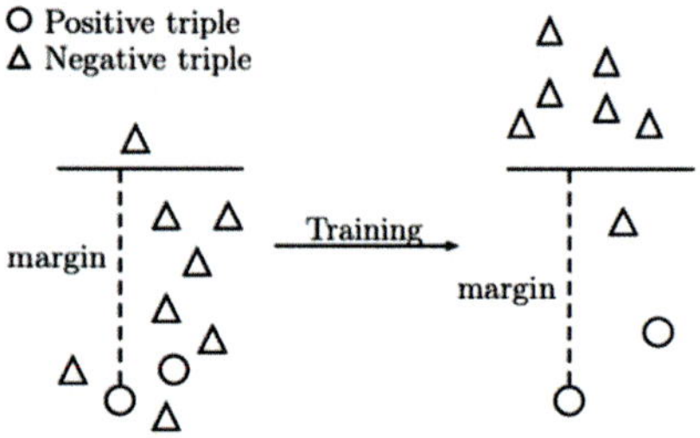

Fig. 2. The effect of training model by negative sampling

4 Experiments

4.1 Preparing Dataset

The classification results are obtained by triple classification to find the incorrect part of the ATS architecture. This task is to verify if any unobserved triples $(\boldsymbol{e}_i, \boldsymbol{r}_k, \boldsymbol{e}_j)$ are correct. The dataset comes from the National ITS Reference Architecture (ARC-IT 9.0)[1] developed by the U.S. Department of Transportation. The part we use is the exchange process of information flow with a physical view of the ATS architecture, including the source and destination physical objects and the information flow. The physical source object acts as the head entity $\boldsymbol{e}_i$, the destination physical object acts as the tail entity $\boldsymbol{e}_j$, and we take the information flow as the relation $\boldsymbol{r}_k$. The dataset is respectively divided into a training dataset, a validation dataset, and a test dataset for model training, parameter tuning, and evaluation. The training dataset contains 1896 positive triples, 154 entities and 905 relations. We constructed 14 negative samples for every positive triple (Table 1).

[1] The Architecture Reference for Cooperative and Intelligent Transportation (ARC-IT) provides a common framework for planning, defining, and integrating intelligent transportation systems. ARC-IT 9.0 includes all views of the National ITS Reference Architecture - Enterprise, Functional, Physical and Communications views. The information is available at https://www.arc-it.net/index.html.

Table 1. A sample for training the architecture embedding model

Sources (head entity)	Flow (relation)	Destination (tail entity)
Alerting and advisory system	Alerts and advisories	Emergency management center
Basic vehicle	Driver input information	Vehicle OBE
Basic vehicle	Host vehicle status	Vehicle OBE
Connected vehicle roadside equipment	Intersection status	Commercial vehicle OBE
Connected vehicle roadside equipment	Signal priority status	Commercial vehicle OBE
Connected vehicle roadside equipment	Data provision	Data distribution system
Connected vehicle roadside equipment	Data query	Data distribution system
...	...	...

4.2 Evaluation Results

We set up an evaluation scheme similar to the TransE model [17]. We first generate test data for evaluation. For each positive triple in the test or validation set, we construct 10 negative triples by randomly replacing entities, five in the head position and the other five in the tail position. To make the evaluation process as accurate as possible, we use only the entities that occurred in that position to replace the corresponding position, and further ensure that the replaced triples do not exist in the training, validation or test datasets. We only use every triple's truth value as the classifying criterion. Triples with large truth values are often predicted to be correct. F_1 is used here to measure the accuracy of the triple classification task. As shown in Table 2, let TP be the triples that our method correctly predicts to hold, let FP be the triples that our method incorrectly predicts to hold, and FN be the triples that our method incorrectly predicts not to hold. The F_1-score formula is established as follows:

Table 2. Results of triple classification

Predicted	True triple	
	Positive	Negative
Positive	TP	FP
Negative	FN	TN

$$\text{Precision} = \frac{TP}{TP + FP} \tag{3}$$

$$\text{Recall} = \frac{TP}{TP + FN} \tag{4}$$

$$F_1 = 2 \times \frac{\text{Precision} \cdot \text{Recall}}{\text{Precision} + \text{Recall}} \tag{5}$$

We set up multiple training sets with different proportions of triples, using 30%, 50% and 75% of the triples as training sets to calculate the accuracy of the triple classification task (Table 3).

Table 3. The F_1-score accuracy comparison results of triple classification task (in percentage)

Training dataset	Precision(%)	Recall(%)	F_1(%)
30%	74.8	71.3	73.0
50%	76.2	73.5	74.8
75%	80.4	78.2	79.3

The experimental results show that the model gains relatively good results with different sizes of training sets, and the classification accuracy can reach 79.3% as the training set increases, which is positive for separating the incorrect triples from the ATS architecture.

5 Conclusions and Future Work

In this paper, we propose a new method to achieve the purpose of the ATS architecture abnormality detection. We generate the vector representation of the ATS architecture structural features through an embedding model, and classify features of the incorrect type from the ATS architecture by vector computation and triple classification. The experimental results show that the method can diagnose most of the structural errors and achieve good detection accuracy. The method can diagnose ATS architecture without prior knowledge or rules. It has potential for further applications such as tracing and solving problems of the ATS architecture.

For future work, we would like to consider expanding the diagnosing scope of ATS architecture and adding more features of the ATS architecture as embedding content, such as the logic of collaboration between functional objects, the hierarchical relationship between transport services and functional objects, etc. The relationship of "1-to-many" and "many-to-many" between them, grows the complexity of the embedding model and the difficulty of model training. We will also improve the diagnosis framework by integrating rules with the embedding model, and use the improved embedding model to obtain the vector representation of rules, which may enhance the accuracy or efficiency of detection.

Acknowledgments. This study was supported by the National Key R&D Program of China (No. 2020YFB1600400).

References

1. Zhang, J., Wang, F., Wang, K., Lin, W., Xu, X., Chen, C.: Data-driven intelligent transportation systems: a survey. IEEE Trans. Intell. Transp. Syst. **12**(4), 1624–1639 (2011)
2. Lin, J., Yu, W., Zhang, N., Yang, X., Zhang, H., Zhao, W.: A survey on internet of things: architecture, enabling technologies, security and privacy, and applications. IEEE Internet Things J. **4**(5), 1125–1142 (2017)
3. Duan, X., Jiang, H., Tian, D., Zou, T., Zhou, J., Cao, Y.: V2I based environment perception for autonomous vehicles at intersections. China Commun. **18**(7), 1–12 (2021)
4. Fünfrocken, M., Otte, A., Vogt, J., Wolniak, N., Wieker, H.: Assessment of ITS architectures. IET Intel. Transport Syst. **12**(9), 1096–1102 (2018)
5. Vogt, T., et al.: A comprehensive risk management approach to information security in intelligent transport systems. SAE Int. J. Transp. Cyber. Privacy **4**(1), 39–58 (2021)
6. Gao, Z., Cecati, C., Ding, S.X.: A survey of fault diagnosis and fault-tolerant techniques—part I: Fault diagnosis with model-based and signal-based approaches. IEEE Trans. Industr. Electron. **62**(6), 3757–3767 (2015)
7. Li, W., Li, H., Gu, S., Chen, T.: Process fault diagnosis with model- and knowledge-based approaches: advances and opportunities. Control. Eng. Pract. **105**, 104637 (2020)
8. Xiaojun, C., Shengbin, J., Yang, X.: A review: Knowledge reasoning over knowledge graph. Expert Syst. Appl. **141**, 112948 (2020)
9. Chen, G., Liu, M., Kong, Z.: Temporal-Logic-Based semantic fault diagnosis with time-series data from industrial internet of things. IEEE Trans. Industr. Electron. **68**(5), 4393–4403 (2021)
10. Guang, C., Tonghai, J., Meng, W., Xinyu, T., Wenfei, J.: Modeling and reasoning of IoT architecture in semantic ontology dimension. Comput. Commun. **153**, 580–594 (2020)
11. Omran, P.G., Wang, K., Wang, Z.: An embedding-based approach to rule learning in knowledge graphs. IEEE Trans. Knowl. Data Eng. **33**(4), 1348–1359 (2021)
12. Shihong, E.H., Yiheng, F., Henry, X.L.: A data-driven method for falsified vehicle trajectory identification by anomaly detection. Transp. Res. Part C Emerg. Technol. **128**, 103196 (2021)
13. Bordes, A., Usunier, N., Garcia-Durán, A., Weston, J., Yakhnenko, O.: Translating embeddings for modeling multi-relational data. In: Proceedings of the 26th International Conference on Neural Information Processing Systems (NIPS 2013), vol. 2, pp. 2787–2795. Curran Associates Inc., Red Hook (2013)
14. S Suresh, S., Neville, J.: A hybrid model for learning embeddings and logical rules simultaneously from knowledge graphs. In: 2020 IEEE International Conference on Data Mining (ICDM), 17–20 November 2020, pp. 1280–1285 (2020)
15. Wang, Q., Mao, Z., Wang, B., Guo, L.: Knowledge graph embedding: a survey of approaches and applications. IEEE Trans. Knowl. Data Eng. **29**(12), 2724–2743 (2017)
16. Guo, S., Wang, Q., Wang, L., Wang, B., Guo, L.: Jointly embedding knowledge graphs and logical rules. In: Proceedings of the 2016 conference on empirical methods in natural language processing, pp. 192–202 (2016)
17. Lv, X., Hou, L., Li, J., Liu, Z.: Differentiating concepts and instances for knowledge graph embedding. arXiv preprint arXiv:1811.04588 (2018)

Check for updates

Dynamic Electric Bus Control Method for the Route with Dedicated Bus Lane

Yuting Ji, Jinhua Ji, and Yiming Bie(✉)

School of Transportation, Jilin University, Changchun 130022, Jilin, China
yimingbie@126.com

Abstract. Electric buses have the advantages of small noise, zero emission and simple control, which can effectively reduce urban pollutant emissions and energy consumption. Therefore, vigorously promoting the development of electric buses is of great significance to accelerate the low-carbon development of the city and realizes the goal of "carbon peak" and "carbon neutrality". The operating condition of electric buses is an important factor affecting their energy consumption. Ensure buses under reasonable working conditions can improve the operation efficiency, reduce the operation energy consumption and the operation cost of bus enterprises. For the bus route with dedicated bus lane, we divide road sections based on road characteristics and analyze the operation state of buses on different units. Considering the constraints such as the bus travel punctuality rate between stations and the intersection delay rate, we taking the minimum operation energy consumption between stations as the optimization objective. Taking the traveling speed of the bus on the road section unit between stations and the starting moment of green light on the intersection as the optimization variables, the dynamic control model of the electric buses is established. Finally, the simulated annealing algorithm is used to solve the built model. Comparing the optimization scheme with the original scheme target value and analyzing the model optimization effect, the numerical results show that the total operating energy consumption can be saved up to 8.29%, which proves the optimal travel can meet the needs of passengers while reducing operating energy consumption.

Keywords: Electric bus · Dynamic control · Energy consumption · Simulated annealing algorithm

1 Introduction

As an important part of urban transportation, electric buses have the advantages of small noise, high travel stability, zero emission and simple control. It is also a vital measure to solve urban congestion and reduce traffic pollution. Operation energy consumption of electric buses is not only related to the operating mileage, traveling time and the number of passengers, but also closely related to the speed and acceleration of the electric buses. The working conditions of electric buses are important factor affecting energy consumption [1–3]. Ensure electric buses working under reasonable operating conditions can reduce the bus energy consumption, improve the operation efficiency, and

Y. Bie et al. (Eds.): KES-STS 2022, SIST 304, pp. 94–103, 2022.
https://doi.org/10.1007/978-981-19-2813-0_10

reduce the operation cost of bus enterprises. Therefore, how to correctly estimate and reduce the operating energy consumption of electric buses is one of the wide concerns of scholars. The current research on electric buses energy consumption estimation can be divided into kinetic methods and data-driven methods [4, 5]. The former uses the vehicle operating state parameters per second to calculate traction energy consumption and power recovery energy, the latter mainly uses laboratory data or actual observation data to establish a training model of operating energy consumption based on machine learning method.

The control research of electric buses mainly focuses on travel energy consumption [6–11] and vehicle control optimization [12–17], but most of the studies are independently conducted for the above two directions, which lack of discussion and research on the overall operation of the bus. Moreover, the actual operating environment of electric buses is a dynamic system with many uncertainties. The static vehicle control method cannot well adapt to the actual travel environment of bus operation. Therefore, how to apply the existing technology combined with data to improve the travel efficiency, to reduce the energy loss, and to improve the service quality of electric buses is a very important problem for the development of the current urban public transportation system.

For the bus route with dedicated bus lane, the state of the electric bus is relatively simple and easy to control. Therefore, we dynamically optimize the driving parameters of the electric bus and the signal timing parameters of the intersection according to the information such as the position and speed of the bus, so that the bus travels in the best state. Compared with the traditional electric bus control research, we combine the optimization of intersection signal timing and control the bus driving plan from the perspective of the overall operation of the bus, so as to achieve the purpose of energy saving and consumption reduction. Remainder of this paper is organized as follows. Section 2 presents the concept of the dynamic control method. Next, we conduct validation examples to demonstrate the effectiveness of the proposed model and provide the concluding remarks in Sect. 3.

2 Methodology

2.1 Problem Analysis

There are N_1 bus stations and N_2 intersections on the line. The road on the line is divided into intersection units and section units. The intersection units *Int* is the road from the maximum queue length of vehicles on the intersection to the stop line on the intersection. The remainder is divided into section units *Seg*. The driving process of the bus between station i and $i+1$ is from the station i where the bus starts at t_i^s to the station $i+1$ where the bus arrives and waits for passengers to get on and off then drives off. In order to calculate the total energy consumption $E_{i,i+1}$ of all units between station i and $i+1$, we establish an energy optimization model between adjacent sites which takes the minimum $E_{i,i+1}$ as optimization objective and takes the driving speed $V_{i,j}^{Seg}$ and the green light starting moment $g_{i,k}^{s}$ corresponding to the phase m of the intersection unit $Int_{i,k}$ as the optimization variables. The energy optimization model is as follows.

$$\min\ E_{i,i+1} \tag{1}$$

$$s.t.\quad t_{i+1}^{s*} - \varepsilon \le t_{i+1}^{s} \le t_{i+1}^{s*} + \varepsilon \tag{2}$$

$$V_i^{\min} \le V_{i,j}^{Seg} \le V_i^{\max}\, j = 1, 2, 3 \ldots N_i + 1 \tag{3}$$

$$\left| \frac{De_{i,k} - De_{i,k}^{0}}{De_{i,k}^{0}} \right| \le \delta\, k = 1, 2, 3 \ldots N_i \tag{4}$$

where t_{i+1}^{s*} is the departure time at station $i + 1$ specified in the departure schedule; ε is the allowable error time of departure time, s; $V_i^{\min}$ and $V_i^{\max}$ are the minimum and maximum speed limits of the bus at station i and $i + 1$, m/s; $De_{i,k}^{0}$ and $De_{i,k}$ are the average delay time of intersection before and after changing the timing scheme, which can be calculated through delay model; δ is the delay rate of change on the intersection. The constraint (2) indicates that the buses meet the departure time limit of the departure schedule, the departure time of station $i + 1$ is equal to the sum of the departure time of station i and travel time between station i and $i + 1$; The constraint (3) indicates that the driving speed $V_{i,j}^{Seg}$ of the bus between stations i and $i + 1$ will be within the allowable speed range of the section. The constraint (4) indicates that the average delay of the intersection needs to be within the allowable range after adjusting the green light.

2.2 Calculation of Travel Time of the Electric Bus

(i) Bus travel time calculation on the section unit

The bus changes the speed at the initial speed $V_{i,j}^{Seg}(0)$, acceleration a_i at the beginning of the section unit $Seg_{i,j}$ through $t_{i,j}^{Seg}(1)$ to $V_{i,j}^{Seg}$, then keeps the uniform speed for $t_{i,j}^{Seg}$ seconds and reaches the end of the section at the acceleration a_{i2} through $t_{i,j}^{Seg}(2)$. Define the 0–1 variable $Y_{i,j}^{Seg}$, $Y_{i,j}^{Seg} = 0$ indicates the bus accelerating on the unit, or that the bus slowing down on the unit, therefore when $Y_{i,j}^{Seg} = 0$, $a_i = a_{i1}$; $Y_{i,j}^{Seg} = 1$, $a_i = a_{i2}$.

The initial speed of the electric bus at unit $Seg_{i,j}$ between section i and $i + 1$ is expressed as follows:

$$V_{i,j}^{Seg}(0) = \begin{cases} 0 & \textit{The road unit starts with the station or the stop line and } X_i^k = 1 \\ V_{i,k}^{Int} & \textit{The road unit starts with the station or the stop line and } X_i^k = 0 \end{cases} \tag{5}$$

where X_i^k is a 0–1 variable, indicating whether the bus needs to stop on the intersection unit $Int_{i,k}$. If $X_i^k = 0$ the bus does not stop on the intersection unit; otherwise $X_i^k = 1$; $V_{i,k}^{Int}$ is the speed when the bus leaves the intersection k, m/s.

$t_{i,j}^{Seg}(1)$ is calculated by the following equation

$$t_{i,j}^{Seg}(1) = \frac{V_{i,j}^{Seg} - V_{i,j}^{Seg}(0)}{a_i} \tag{6}$$

The calculation of $t_{i,j}^{Seg}(2)$ is affected by the end of the section:

$$t_{i,j}^{Seg}(2) = \begin{cases} \dfrac{V_{i,j}^{Seg}}{|a_{i2}|} & \textit{The section unit ends with stop } i+1 \\ 0 & \textit{The section unit ends with the stop line} \end{cases} \quad (7)$$

The travel time $T_{i,j}^{Seg}$ of the electric bus on the section unit j can be calculated as follows:

$$T_{i,j}^{Seg} = t_{i,j}^{Seg}(1) + t_{i,j}^{Seg} + t_{i,j}^{Seg}(2) \quad (8)$$

(ii) Bus travel time calculation on the intersection unit

When the electric bus from the section unit $Seg_{i,J}$ enters the intersection unit $Int_{i,k}$ with speed $V_{i,J}^{Seg}$, time $t_{i,k}^{s}$. It reaches the intersection stop line at $t_{i,k}^{e}$ through $t_{i,k}^{t}$. When the bus enters the intersection unit, the signal intersection performs the cycle $c_{i,k}$, and the cycle duration is $C_{i,k}$, which starts at $c_{i,k}^{s}$ and ends at $c_{i,k}^{e}$. The cycle has $\varphi_{i,k}$ phases, each phase's green light time is $g_{i,k}^{h}$ $(h \in [1, \varphi_{i,k}])$ and yellow light time is $y_{i,k}^{h}$. The bus is controlled by the phase m $(1 \leq m \leq \varphi_{i,k})$. Phase m starts green light at moment $g_{i,k}^{s}(0)$, at $g_{i,k}^{s}$ after optimization.

$$t_{i,k}^{s} = t_{i}^{s} + \sum_{j=1}^{J} T_{i,j}^{Seg} + \sum_{k=1}^{J-1} T_{i,k}^{Int} \quad (9)$$

The following equation shows the travel time and travel distance when the bus passes through the intersection in different ways:

a. The electric bus passes freely through the intersection

When the bus passes freely, the travel speed $V_{i.k}^{Int}$ of the bus is equal to the section speed $V_{i,J}^{Seg}$ when the bus enters the intersection unit. At this time, the travel time $t_{i,k}^{t}$ of the bus on the intersection unit is equal to the uniform speed travel time $t_{i,k}^{Int}$.

$$t_{i,k}^{Int} = \frac{d_{i,k}^{Int}}{V_{i,k}^{Int}} \quad (10)$$

b. The electric bus speed changes through the intersection

The bus speed changes through the intersection, including the bus accelerates and the bus slows down through the intersection. At this time, the bus needs to leave the intersection at $t_{i,k}^{s}$ with initial speed $V_{i,J}^{Seg}$, acceleration a_i through $t_{i,k}^{Int}(1)$. Define the 0–1 variable $Y_{i,j}^{Seg}$, when $Y_{i,j}^{Int} = 0, a_i = a_{i1}$; $Y_{i,j}^{Int} = 1, a_i = a_{i2}$.

$$t_{i,k}^{Int}(1) = \sqrt{\frac{2x_{i,k}^{Int}}{|a_i|}} \tag{11}$$

The speed change time $t_{i,k}^{Int}(1)$ of the bus is determined by the length $x_{i,k}^{Int}$ of the intersection unit. However, the bus travel on the line is limited by the road conditions, so when the bus accelerates to the maximum speed allowed by the section, the bus stops accelerating and keeps the speed through the intersection. The deceleration case has same argument. At this time the $t_{i,k}^{Int}(1)$ is:

$$t_{i,k}^{Int}(1) = \frac{V_i^{'} - V_{i,J}^{Seg}}{a_i} \tag{12}$$

where $V_i^{'}$ is the speed limit of the bus between station i and $i+1$, when accelerating $V_i^{'} = V_i^{\max}$, otherwise, $V_i^{'} = V_i^{\min}$, m/s.

The travel time $t_{i,k}^{t}$ of the bus is expressed as follows:

$$t_{i,k}^{t} = t_{i,k}^{Int}(1) + t_{i,k}^{Int} \tag{13}$$

c. The electric bus cannot directly pass through the intersection

When the electric bus cannot directly pass the intersection, the bus needs to keep the speed of the starting line of intersection unit, and slows down to 0 at the stop line of the intersection at acceleration a_{i2}. The deceleration time $t_{i,k}^{t-}$, and the bus travel time $t_{i,k}^{t}$ are the same as Eq. (12)–(13). Among, $V_i^{'} = 0,\ \ a_i = a_{i2}$.

The time $t_{i,k}^{wait}$ of the bus waiting on the intersection is equal to:

$$t_{i,k}^{wait} = g_{i,k}^{s} - t_{i,k}^{e} \tag{14}$$

Above all, the travel time $T_{i,k}^{Int}$ of the electric bus on the intersection unit between stations i and $i+1$ is expressed as follows:

$$T_{i,k}^{Int} = \begin{cases} t_{i,k}^{t} + t_{i,k}^{wait} & X_i^k = 1 \\ t_{i,k}^{t} & X_i^k = 0 \end{cases} \tag{15}$$

The total travel time of the electric bus between stations i and $i+1$ is equal to the sum of the travel time on all road units and intersection units between stations i and $i+1$.

2.3 Energy Consumption Calculation of the Electric Bus

The operating consumption $E_C(t_C)$ of the electric bus during t_C can be obtained by the instantaneous power $P_W(t)$ on the wheel:

$$P_W(t) = (ma_i + mg\cos(\theta)\frac{f(c_1V(t) + c_2)}{1000} + \frac{1}{2}\rho_{air}AC_DV^2(t) + mg\sin(\theta))V(t) \tag{16}$$

where m is the bus weight, kg; g is the gravitational acceleration, m/s^2; f is the rolling resistance coefficient; θ is the road slope angle, rad; c_1, c_2 is the rolling resistance parameter; ρ_{air} is air density; i is road slope, kg/m^3; A is automobile windward area, m^2; C_D is air resistance coefficient; $V(t)$ is the travel speed of the vehicle at the fixed speed (details from 2.2), which can be calculated when the bus is in accelerated state, m/s.

$$V(t) = V(0) + a_i t \tag{17}$$

where $V(0)$ is the initial speed when the bus starts accelerating (detailed calculation methods from 2.2), m/s; $a_i = a_{i1}$.

$$E_C(t_C) = \int_0^{t_C} \frac{P_W(t)}{\eta_1\eta_2\eta_3} dt \tag{18}$$

where η_1 is transmission system efficiency; η_2 is power generation efficiency for motor; η_3 is charging efficiency for battery pack.

In addition, when the motor is in the generator state under the wheel inertia, the power output is negative, and the energy generated by the brake inch can be recovered by the recovery device. The braking energy $E_R(t_R)$ of the vehicle during deceleration t_R seconds is expressed as follows:

$$E_R(t_R) = \int_0^{t_R} -P_W(t)\eta_1\eta_2\eta_3\eta_4 dt \tag{19}$$

$$\eta_4 = e^{-\frac{\partial}{a_i}} \tag{20}$$

where η_4 is regenerative braking efficiency; $a_i = a_{i2}$.

Therefore, the operation energy consumption $E_{i,j}^{Seg}$ of the bus section unit and the operation energy consumption $E_{i,k}^{Int}$ of the bus section unit between the stations i and $i+1$ are expressed as follows:

$$E_{i,j}^{Seg} = \begin{cases} E_C(t_{i,j}^{Seg}(0)) + E_C(t_{i,j}^{Seg}) - E_R(t_{i,j}^{Seg}(1)) & Y_{i,j}^{Seg} = 0 \\ E_C(t_{i,j}^{Seg}) - E_R(t_{i,j}^{Seg}(0)) - E_R(t_{i,j}^{Seg}(1)) & Y_{i,j}^{Seg} = 1 \end{cases} \tag{21}$$

$$E_{i,k}^{Int} = \begin{cases} E_C(t_{i,k}^{Int}) + E_C(t_{i,k}^{Int}(1)) & Y_{i,k}^{Int} = 0 \\ E_C(t_{i,k}^{Int}) - E_R(t_{i,k}^{Int}(1)) & Y_{i,k}^{Int} = 1 \end{cases} \tag{22}$$

The total energy consumption $E_{i,i+1}$ of the electric bus between stations i and $i+1$ is expressed as follows:

$$E_{i,i+1} = \sum_{j=1}^{N_i+1} E_{i,j}^{Seg} + \sum_{k=1}^{N_i} E_{i,k}^{Int} \tag{23}$$

3 Solution Algorithm

Simulated annealing is a kind of heuristic algorithm, which is widely used because of its advantages of simple description, high efficiency and less constraint of initial conditions. In this paper, the travel speed of the bus and the starting moment of green light on the intersection affect each other in time, and the best starting moment cannot be obtained through the determined speed optimization, which increases the difficulty of solving the model to some extent. Therefore, we select the simulated annealing algorithm, which applies the analogy of solid matter in physics to the optimization model, taking the actual speed of the bus and the signal timing scheme on the intersection as the initial solution, and gradually adjusting to solve the optimal travel scheme of the vehicle on the bus line.

Step 1: Set the initial parameters. Set the starting temperature $T_0 = 120$ °C, termination temperature $T_f = 1$ °C, temperature attenuation coefficient $\gamma = 0.99$, number of perturbations at each temperature $Markov = 100$. Determine the initial state $T = T_0$, set the actual running data as the initial solution $plan_i = [V_{i,1}^{Seg}, V_{i,2}^{Seg}, \ldots, V_{i,N_i}^{Seg}, g_{i,1}^{Int}, g_{i,2}^{Int}, \ldots, g_{i,N_i}^{Int}]$, then take it as the optimal solution $plan\_best$. Calculate the energy consumption of the initial solution $energy_i$, and take it as the optimal target value $energy\_best$.

Step 2: Produce a new solution. If $T > T_f$, produce a new solution $plan\_new = plan_i$ and generate the random number *rand*, by which changes the traveling speed of the bus on the section unit between stations and the starting moment of green light on the intersection.

Step 3: Select solution. Calculate the target value of the new solution, comparing the initial solution with the new solution to determine whether the new solution is accepted. If the target value under the optimization scheme exceeds the corresponding target value of the initial solution, the generated optimization solution is replaced with the optimal solution. Conversely, compare the generated random number with the cooling probability. If the random number is less than the cooling probability, accept the new solution and replace the optimal solution, otherwise, the optimal solution remains unchanged.

Step 4: Repeat the perturbation. At this temperature, Step2 and Step3 are repeated *Markov* times.

Step 5: Decay temperature. Repeat Step2–Step4 at the new temperature until the temperature is equal to the end temperature T_f.

Step 6: End. Output the optimal scheme solution from the recording and the corresponding minimum energy consumption.

4 Case Study

4.1 Data Description

We collect the data of a bus from line 108 in a city, which has 25 stations. The total length of the bus line is 8.1 km and there are four signalized intersections on it. Under normal circumstances, the whole trip takes 1457 s and 3.5 kWh. For the sake of description, sections are numbered from 1 to 24 and intersections are numbered from 1 to 4. The bus trip starts at 10:36:09 and ends at 11:00:26. The traveling time requirements of the line are as follows: Section 1 → Section 3 (4 min) → Section 7 (10 min) → Section 11 (15 min) → Section 19 (23 min) → Section 24 (30 min).

Above all, the departure moments of the bus at the stations are the time constraint control points of the trip, that is, the bus must finish Section 3 before 10:40:09, Section 7 before 10:46:09, Section 11 before 10:51:09, Section 19 before 10:59:09, Section 24 before 11:06:09. And the average acceleration of the electric bus is set as $a_{i1} = 0.51$ m/s^2 and $a_{i2} = -0.48$ m/s^2. Other parameters are expressed as follows (Table 1):

Table 1. Parameter table

Parameters	Values	Parameters	Values
η_1	0.9	C_D	0.7
η_2	0.9	δ	1.21
η_3	0.99	i_0	0
∂	0.0411	A	8.037
c_1	0.0328	m	8275
c_2	4.575	V_i^{min}	5.6
f	0.012	V_i^{max}	16.6

4.2 Results Analysis

Taking the actual operation solution as the initial solution, the travel scheme brought into the model optimization calculation is shown in Table 2. In the optimization scheme, the bus starts at 10:36:09 and finishes Section 24 at 11:00:33, taking 1504 s, which is within the travel time limit of 30 min ensuring the demand of punctuality and meeting the needs of passengers. The green light at Intersection 3 will start 10 s in advance, and the signal timing scheme at other intersections will not change. Besides, under the optimization scheme, the travel energy consumption between the corresponding stations decreased from 0.21 kWh, 0.15 kWh, 0.05 kWh, 0.12 kWh to 0.16 kWh, 0.14 kWh, 0.03 kWh and 0.11 kWh respectively. The whole line consumption is 3.21 kWh. Compared with the 3.5 kWh of the original scheme, it decreased by 8.29%, which shows that optimizing the travel scheme of the electric bus based on the number of bus passengers can reduce the operation energy consumption of single-line buses, and save bus operation cost, which further proves the effectiveness of the built model.

Table 2. Travel scheme

Section number	Leaving time	Arrival time	Traveling speed (m/s)	Section number	Leaving time	Arrival time	Traveling speed (m/s)
1	10:36:09	10:37:29	6.53	13	10:49:28	10:50:17	8.77
2	10:37:30	10:38:03	8.15	14	10:50:20	10:51:11	8.55
3	10:38:03	10:39:05	7.15	15	10:51:11	10:52:10	4.04–0–8.43
4	10:39:07	10:40:22	12.15	16	10:52:29	10:53:20	9.05
5	10:40:35	10:41:20	9.15	17	10:53:23	10:54:08	8.59
6	10:41:25	10:43:04	5.8–12.5	18	10:54:10	10:55:03	10.49
7	10:43:05	10:43:47	10.15	19	10:55:05	10:55:37	8.5
8	10:44:10	10:45:10	10.85	20	10:55:39	10:56:36	9.51
9	10:45:12	10:45:55	9.95	21	10:56:36	10:57:26	10.60
10	10:45:56	10:46:49	11.85	22	10:57:30	10:59:04	3.01–0–9.11
11	10:46:51	10:47:51	5.97–9.67	23	10:59:05	10:59:58	9.69
12	10:48:16	10:49:25	10.75	24	10:59:59	11:01:13	9.58

5 Conclusions

In this paper, for the bus route with dedicated bus lane, we analyze the movement of the electric bus between adjacent stations. With the optimization objective of minimum energy consumption between stations, taking the operation speed of the bus on the road section unit between stations and the starting moment of green light on the intersection as the optimization variables, the bus travel punctuality rate between stations and the intersection delay rate as the constraints, the dynamic control model of the electric bus is established. The model is solved with the actual survey data. Results show that the total energy consumption is 3.21 kWh, which is reduced by 0.29 kWh compared with the original energy consumption, accounting for 8.29%. It shows that this model can effectively reduce the energy consumption of bus operation, proving its effectiveness and feasibility. However, the optimization model designed in this paper takes only into account the dynamic control of single line single bus, and does not consider the influence of other buses on the same line and buses on different lines. In the future, the above factors can be considered to build the dynamic control method of electric buses.

Acknowledgments. This study was supported by the National Natural Science Foundation of China (No. 71771062), China Postdoctoral Science Foundation (No. 2019M661214 & 2020T130240), and Fundamental Research Funds for the Central Universities (No. 2020-JCXK-40).

References

1. Kivekäs, K., Vepsäläinen, J., Tammi, K.: Stochastic driving cycle synthesis for analyzing the energy consumption of a battery electric bus. IEEE Access **6**, 55586–55598 (2018)
2. Nader, A.E., Aboelsood, Z., Hany, E.Z.F.: Novel electric bus energy consumption model based on probabilistic synthetic speed profile integrated with HVAC. IEEE Trans. Intell. Transp. Syst., 1–15 (2020)
3. Qin, K., Liu, Y., Hu, X.: Variable parameter self-adaptive control strategy based on driving condition identification for plug-in hybrid electric bus. J. Beijing Inst. Technol. **28**(1), 162–170 (2019)
4. Li, P., Zhang, Y., Zhang, Y., Zhang, K.: Prediction of electric bus energy consumption with stochastic speed profile generation modelling and data driven method based on real-world big data. Appl. Energy **298** (2021)
5. Zhang, J., Tang, T., Yan, Y., Qu, X.: Eco-driving control for connected and automated electric vehicles at signalized intersections with wireless charging. Appl. Energy **282**, 116215 (2021)
6. Yang, R., Yang, X., Huang, W., Zhang, S.: Energy management of the power-split hybrid electric city bus based on the stochastic model predictive control. IEEE Access **9**, 2055–2071 (2021)
7. Kotiev, G.O., Butarovich, D.O., Kositsyn, B.B.: Energy efficient motion control of the electric bus on route. IOP Conf. Ser. Mater. Sci. Eng. **315**(1), 012014 (2018)
8. Gao, K., Yang, Y., Sun, L., Qu, X.: Revealing psychological inertia in mode shift behaviorand its quantitative influences on commuting trips. Transp. Res. F Traffic Psychol. Behav. **71**, 272–287 (2020)
9. Peled, I., Lee, K., Jiang, Y., Dauwels, J., Pereira, F.C.: On the quality requirements of demand prediction for dynamic public transport. Commun. Transp. Res. **1**, 100008 (2021). https://doi.org/10.1016/j.commtr.2021.100008
10. Luo, G., Luo, Y., Li, K.: Control strategy for plug-in hybrid electric bus based on optimal electric energy use. Automot. Eng. **34**(6), 475–478 (2012)
11. Peng, B., Keskin, M.F., Kulcsár, B., Wymeersch, H.: Connected autonomous vehicles for improving mixed traffic efficiency in unsignalized intersections with deep reinforcement learning. Commun. Transp. Res. **1**, 100017 (2021)
12. Tian, H., Li, S., Wang, X., Huang, Y., Tian, G.: Data-driven hierarchical control for online energy management of plug_in hybrid electric city bus. Energy **142**, 55–67 (2018)
13. Hao, M., Bie, Y., Zhang, L., Mao, C.: Improving schedule adherence based on dynamic signal control and speed guidance in connected bus system. J. Intell. Connect. Veh. **3**(2), 79–88 (2020). https://doi.org/10.1108/JICV-06-2020-0005
14. Lu, C., Liu, C.: Ecological control strategy for cooperative autonomous vehicle in mixed traffic considering linear stability. J. Intell. Connect. Veh. **4**(3), 115–124 (2021). https://doi.org/10.1108/JICV-08-2021-0012
15. Zhang, W., Zhao, H., Xu, M.: Optimal operating strategy of short turning lines for the battery electric bus system. Commun. Transp. Res. **1**, 100023 (2021). https://doi.org/10.1016/j.commtr.2021.100023
16. Bie, Y., Ji, J., Wang, X., Qu, X.: Optimization of electric bus scheduling considering stochastic volatilities in trip travel time and energy consumption. Comput. Aided Civil Infrast. Eng. **36**(12), 1530–1538 (2021)
17. Bie, Y., Hao, M., Guo, M.: Optimal electric bus scheduling based on the combination of all-stop and short-turning strategies. Sustainability **13**(4), 1827 (2021)

Check for updates

Evaluating the Impact of Signal Control on Emissions at Intersections

Jieyu Fan[1(✉)], Martin Baumann[1], Sarang Jokhio[1], and Jie Zhu[2]

[1] Faculty of Engineering, Computer Science and Psychology, Department of Human Factors, Ulm University, 89069 Ulm, Germany
fan.jieyu@uni-ulm.de

[2] Department of Architecture and Civil Engineering, Chalmers University of Technology, 412 96 Goteborg, Sweden

Abstract. Transport emission has become an increasingly serious problem, and it is an urgent issue in sustainable transport. In this study, by constructing traffic emission models for different vehicle types and operating conditions, the changes in CO, HC, and NOx emissions of light-duty and heavy-duty vehicles before and after signal control optimization were quantified based on VISSIM simulation. The OBEAS-3000 vehicle emission testing device was used to collect data on the micro-operational characteristics of different vehicles under different operating conditions as well as traffic emission data. Based on the data collected, the VSP (Vehicle Specific Power) model combined with the VISSIM traffic simulation platform was used to calculate the emissions of light and heavy vehicles in the mixed traffic flow before and after intersection signal optimization. It is known from the study that signal control optimization has a greater impact on heavy vehicles than on light vehicles. Emissions of CO, HC, and NOx from heavy vehicles and light vehicles are all reduced, but NOx emissions from light vehicles remain largely unchanged. The research results reveal the emission patterns of light and heavy vehicles in different micro-operating conditions and establish a traffic emission model. It provides a theoretical basis for accurate traffic emission analysis and traffic flow optimization, as well as a scientific basis for the formulation of traffic management measures and emission reduction in large city transport systems.

Keywords: Mixed traffic flow · Instantaneous emissions · Traffic simulation

1 Introduction

In recent years, vehicle ownership in Chinese major cities has increased year by year, it brings convenience to transportation but also causes numerous urban problems, especially urban air pollution caused by traffic emissions [1, 2], and pollutants from traffic emissions have become one of the main sources of urban pollutants [3]. Global environmental problems are becoming increasingly serious, and with the increase of motor vehicles, traffic emissions are one of the "main culprits" [4]. According to the proportion of pollutants emitted by air pollution sources published by the Chinese Environmental

Y. Bie et al. (Eds.): KES-STS 2022, SIST 304, pp. 104–111, 2022.
https://doi.org/10.1007/978-981-19-2813-0_11

Protection Administration, motor vehicle emissions account for 20%–30% of air pollution, and in some cities even reach 30%–50%, urban air pollution is gradually transforming from "soot type" to "tailpipe type" [5, 6]. Excessive emissions of pollutants from transport have already had a serious negative impact on air quality, public health and climate [7]. Vehicles account for 87.7% of traffic emissions of CO, 84.1% of HC and 92.5% of NOx [8]. Vehicle fuel consumption in peak hours increases by an average of 10%, CO, HC, and NOx emissions in-crease by 20% compared to off-peak hours [9].

In mixed urban traffic flows, especially where heavy vehicles account for a certain proportion, the high density of vehicles and the many interweaving points result in high traffic emissions, which not only endanger the urban living environment but also cause incalculable economic losses [10–13]. Traffic emissions have become one of the most important problems facing cities [14]. Urban managers need management measures and instruments that can effectively reduce traffic emissions. This requires accurate analysis and modeling of motor vehicle emission patterns [15]. This study aims to analyze the emission patterns of light and heavy vehicles, a traffic emission model for different vehicle types and operating conditions is construct-ed. On this basis, the emission ratios of different vehicle types in mixed traffic flows are quantified by building traffic simulations.

2 Literature Review

Traffic signals have come a long way since traffic signal control was first introduced to prevent traffic accidents. Kerosene-lit traffic lights were first used as traffic signals in London's Westmeath district in 1868 [16]. The hand-controlled three-colored traffic signal (red, green, and yellow) was first used in 1925 in Piccadilly Street, London, England [17], the yellow light was placed before the appearance of the red light as a preparatory signal for drivers to stop. The advent of intersection signals has improved the transport efficiency of traffic systems, and typically traffic signals operate in three modes: fixed signal timing, prior and adaptive control [18, 19].

In the early stages of traffic signal development, researchers developed methods to determine fixed signal timings assuming that the traffic flow from each intersection remained constant [20], which did not take into account the uncertainty of traffic flow and has lost its relevance in the contemporary traffic climate [21]. From the recognition of the uncertainty of traffic, a great deal of research work has been de-voted to improving the analysis of delay models and the development of computer software [22].

In previous studies, scholars have focused more on optimizing signal timing in terms of vehicle queue lengths, stopping times, and vehicle delays at intersections [23]. But there is no focus on the transport environment. As traffic pollution has become more serious, researchers have not only limited research to solving the problems of time and delays, but more and more studies are linking signal control and traffic emissions. Scientific signal control can be effective in reducing traffic emissions. As early as 1981, Gipps et al. [24] assumed the speed and acceleration of following vehicles based on the desired braking that each driver sets for himself under intersection signal control, the traffic prevalence was assessed to have an impact on traffic emissions. Zhang et al. [25] investigated the relationship between vehicle emissions and simulated operating conditions, traffic emissions were measured by superimposing vehicle emissions under different simulated operating conditions (queuing and waiting at intersections, the proportion of accelerating vehicles, the proportion of decelerating vehicles, etc.) for vehicles at signalized intersections. Meszaros [26] gave traffic flow parameters, such as the concentration of traffic flow and the overall speed of traffic flow, based on real traffic data from the investigated intersections. Not only intersections but also the whole road network were calculated the emissions of CO2, CO, CH, NOx.

In mixed traffic flows, particularly with a proportion of heavy vehicles, it is unclear whether optimizing traffic signal control will achieve emission reductions. Most of the existing studies have been conducted for single traffic flow situations, while less research has been conducted on mixed traffic flows, and the changes in vehicle emissions under different operating conditions have not been considered. This study summarises research and makes a quantitative study of emissions from mixed traffic flows in intersections, and compared the effects of signal control optimization on light and heavy vehicles, and analyzed the changes in CO, HC, and NOx emissions before and after signal optimization.

3 Methods

3.1 VSP Model

In order to obtain more accurate vehicle emission factors and build a detection system for traffic emissions, this study made use of the vehicle driving data obtained by the testing equipment in the urban road. The distribution pattern of motor vehicle emissions was different under different operating conditions of the vehicle. To unify the calibrated traffic emission model and improve the accuracy of traffic emission calculation, this study uniformly adopted VSP distribute intervals to study the emission factors of vehicles. VSP is the instantaneous power per unit mass of the vehicle in kW/t, and the transient emissions of the vehicle are closely related to the VSP value [27]. The formula VSP of is:

$$VSP = v \times (a + g \times grade(\%) + g \cdot C_R) + \frac{1}{2} \cdot \rho_\alpha \cdot \frac{C_D}{m} \cdot A \cdot v^3 \tag{1}$$

where v is the instantaneous speed, m/s; a is the instantaneous acceleration, m/s^2; g is the acceleration of gravity, m/s^2; *grade* is the road gradient, %; C_R is the rolling resistance coefficient; ρa is the air ambient density; C_D is the air resistance coefficient, m^2; m is the total vehicle mass, kg.

Wyatt [28] provided detailed values on VSP modeling based on the data taken for each relevant parameter of the light vehicle. The VSP values are related to speed and acceleration and it can be expressed in Eq. (2).

$$VSP = v \times (1.1a + g \times grade + 0.132) + 0.000302v^3 \tag{2}$$

The VSP formula is not uniform because of the large variability in the values taken for each parameter of heavy vehicle [29]. This study simplified the VSP calculation formula for heavy vehicles using vehicle weight, front-end cross-section, and other parameters to obtain.

$$VSP = v \times (a + g \times grade + 0.09199) + 0.000169v^3 \tag{3}$$

In the formula, the vehicle speed and acceleration are real-time data of the vehicle, the road gradient is 0 and the gravitational acceleration is 9.81 m/s^2.

By making full use of real-time vehicle operating data to accurately quantify the instantaneous emission rate based on VSP, we divided the VSP value by a step of 1 kW/t to generate the BIN partition.

$$\forall VSP \in VSP_{\mathrm{BIN}_i} = \begin{cases} (-\infty, -30] \\ [n-1, n), n = (-29, 29], n \in Z \\ [30, +\infty) \end{cases} \tag{4}$$

3.2 Mixed Traffic Flow VSP

Based on the real-time speed and acceleration recorded by the instrument, the corresponding instantaneous motor vehicle emissions detected by the OBEAS-3000 system are simultaneous. The system can relate instantaneous emission rates to VSP values by averaging multiple instantaneous emission values for the same VSP interval. The mean value of the instantaneous emissions in each interval of the VSP is obtained. These results construct a relationship between light and heavy vehicle VSP intervals and vehicle proportions.

Table 1. Proportion of light and heavy vehicles in different VSP intervals

VSP intervals (kW/t)	Proportion of light vehicles	Proportion of heavy vehicles
[−5, 5]	60.31%	50.26%
[−10, 10]	67.93%	65.37%
[−15, 15]	75.20%	80.11%
[−20, 20]	88.46%	90.62%
[−25, 25]	94.28%	95.81%
[−30, 30]	97.33%	98.45%

4 Case Study and Results

4.1 VISSIM Simulation

In this study, the intersection of Caoan Highway North Jiasong Road in Shanghai was selected as the research object. The traffic volume in VISSIM was input according to the traffic volume of the intersection in the field survey, and the road network was built according to the actual construction of the intersection in the original scenario of the simulation. The VISSIM simulation platform was built based on field research. The road network structure was divided into two parts: road sections and connectors. In the process of constructing road sections, the length of the road section coverage in the simulation was 500m in order to facilitate subsequent lanes and to ensure smooth vehicle operation before entering the intersection in VISSIM, it could reduce vehicle operation errors. The running screen is shown in Fig. 1.

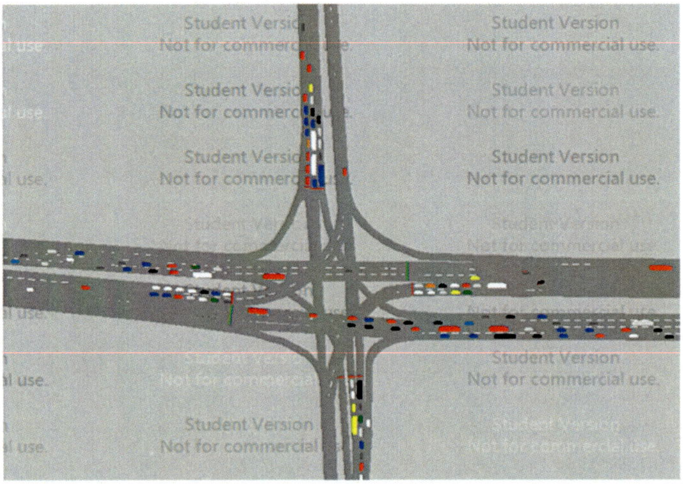

Fig. 1. Simulation screen

4.2 Signal Timing Optimization

The signal timing cycle at the intersection of Caoan Highway North Jiasong Road is 230 s. The timing of each phase's signal cycle is shown in Table 2, where the yellow light flashes for 3 s and the all-red time is 2 s. In Table 2, 1 is east-west straight ahead, 2 is east-west left turn, 3 is north-south straight ahead, 4 is north-south left turn.

The optimum signal period is 205 s according to the headway of heavy ve-hicles and light vehicles. The left-turn signal timing for the east-west import is 29 s, the straight-ahead signal timing is 41 s, the left-turn signal timing for the north-south import is 45 s, and the straight-ahead signal timing is 69 s.

Table 2. Signal timing diagram before and after optimization by conventional method

	1(S)	2(S)	3(S)	4(S)
Red light before signal optimization	177	187	157	177
Yellow light before signal optimization	3	3	3	3
Green light before signal optimization	50	40	70	50
Red light after signal optimization	173	161	157	133
Yellow light after signal optimization	3	3	3	3
Green light after signal optimization	29	41	45	69

4.3 Emission Calculation

For intersection emissions, the peak hourly emissions of CO, HC and NOx for heavy and light vehicles before and after intersection signal control optimization are shown in Table 3.

Table 3. Emissions of light vehicle and heavy vehicle before and after signal control optimization

	Light vehicles before signal optimization (g/h)	Light vehicles after signal optimization (g/h)	Heavy vehicles before signal optimization (g/h)	Heavy vehicles after signal optimization (g/h)
CO	1208.69	1169.92	3129.98	2713.72
HC	196.62	188.33	425.32	361.03
NO_X	32.65	32.53	542.88	473.78

By vehicle type, the CO emissions of the light vehicle in the peak hour were 1208.69 g before signal control optimization and 1169.92 g after signal control optimization, a reduction of 3.21%.

HC emissions were 196.62 g in the peak hour for light vehicles and 188.33 g in the peak hour for light vehicles after signal control optimization, a reduction of 4.22%. The NOx emissions from light vehicles in the peak hour were 32.65 g. After signal control optimization, the NOx emissions from light vehicles in the peak hour were 32.53 g, and the NOx emissions remained unchanged.

CO emissions were 3129.98 g for heavy vehicles in the peak hour before signal control optimization, and 2713.72 g for heavy vehicles in the peak hour after signal control optimization, a reduction of 13.30%. HC emissions were 425.32 g in the peak hour for heavy vehicles, and 361.03 g in the peak hour for heavy vehicles after signal control optimization, a reduction of 15.12%. The NOx emissions of heavy vehicles in the peak hour were 542.88 g. After signal control optimization, the NOx emissions of heavy vehicles in the peak hour were 473.78 g, a reduction of 12.73%.

5 Summary

This study analyses the impact of intersection signals on traffic emissions, and signal control optimization of intersections is carried out based on vehicle conversion factors. VISSIM simulation is used for data collection of vehicle operating conditions to analyse traffic emissions at intersections. The results of their research are as follows.

(1) After signal control optimization, emissions from both heavy and light vehicles in the intersection are reduced, and the effect on emissions from heavy vehicles is more significant than those from light vehicles.
(2) After signal control optimization, the emissions of CO, HC, and NOx of heavy vehicles are reduced, as well as CO and HC of light vehicles, but NOx emissions of light vehicles remained unchanged.

This study quantifies the changes in CO, HC, and NOx emissions from light and heavy vehicles before and after signal control optimization. It can quantify the emission patterns of vehicles under traffic control, providing a theoretical basis for the development of measures to reduce traffic emissions.

References

1. Wang, Q., Yao, Z., Huo, H., He, K.: Emission characteristics of urban light-duty vehicles in China. J. Environ. Sci. **28**(9), 1713–1719 (2008)
2. Li, A., Gao, K., Zhao, P., Qu, X., Axhausen, K.W.: High-resolution assessment of environmental benefits of dockless bike-sharing systems based on transaction data. J. Clean. Prod. **296**, 126423 (2021). https://doi.org/10.1016/j.jclepro.2021.126423
3. Qu, X., Wang, S., Niemeier, D.: On the urban-rural bus transit system with passenger-freight mixed flow. Commun. Transp. Res. **2**, 100054 (2022)
4. Dey, S., Mehta, N.S.: Automobile pollution control using catalysis. Resour. Environ. Sustain. **2**, 100006 (2020)
5. Yu, Y.: Study on the development of motor vehicles and the characteristics of particulate emission pollution in Beijing. Beijing University of Architecture (2015)
6. Gao, K., Yang, Y., Zhang, T., Li, A., Qu, X.: Extrapolation-enhanced model for travel decision making: an ensemble machine learning approach considering behavioral theory. Knowl.-Based Syst. **218**, 106882 (2021). https://doi.org/10.1016/j.knosys.2021.106882
7. Anahita, J., Ioannis, P., Markos, P.: Bart: a mesoscopic integrated urban traffic flow-emission model. Transp. Res. Part C: Emerging Technol. **75**, 45–83 (2017)
8. Huang, Z., Hao, C., Wang, J.: Analysis of vehicle pollutant emissions-Part II of China motor vehicle environmental management annual report (2017). Environ. Prot. **13**, 43–48 (2017)
9. Arti, C., Sharad, G.: Urban real-world driving traffic emissions during interruption and congestion. Transp. Res. Part D: Transp. Environ. **43**, 59–70 (2016)
10. Gao, K., Yang, Y., Qu, X.: Diverging effects of subjective prospect values of uncertain time and money. Commun. Transp. Res. **1**, 100007 (2021). https://doi.org/10.1016/j.commtr.2021.100007
11. Panis, L.I., Broekx, S., Liu, R.: Modelling instantaneous traffic emission and the influence of traffic speed limits. Sci. Total Environ. **371**(1), 270–285 (2006)

12. Gao, K., Yang, Y., Sun, L., Qu, X.: Revealing psychological inertia in mode shift behaviorand its quantitative influences on commuting trips. Transport. Res. F Traffic Psychol. Behav. **71**, 272–287 (2020). https://doi.org/10.1016/j.trf.2020.04.006
13. Chen, K., Yu, L.: Microscopic traffic-emission simulation and case study for evaluation of traffic control strategies. J. Transp. Syst. Eng. Inf. Technol. **7**(1), 93–99 (2007)
14. Srinivasan, K.K., Bhargavi, P.: Longer-term changes in mode choice decisions in Chennai: a comparison between cross-sectional and dynamic models. Transportation **34**(3), 355–374 (2007)
15. Ma, X., Lei, W., Andréasson, I., Chen, H.: An evaluation of microscopic emission models for traffic pollution simulation using on-board measurement. Environ. Model. Assess. **17**(4), 375–387 (2012)
16. Pan, C., Xu, J., Fu, J.: Effect of gender and personality characteristics on the speed tendency based on advanced driving assistance system (ADAS) evaluation. J. Intell. Conn. Veh. **4**(1), 28–37 (2021). https://doi.org/10.1108/JICV-04-2020-0003
17. Mueller, E.A.: Aspects of the history of traffic signals. IEEE Trans. Veh. Technol. **19**(1), 6–17 (1970)
18. Yu, X.H., Recker, W.W.: Stochastic adaptive control model for traffic signal systems. Transp. Res. Part C Emerging Technol. **14**(4), 263–282 (2006)
19. Gradinescu, V., Gorgorin, C., Diaconescu, R., Cristea, V., Iftode, L.: Adaptive traffic lights using car-to-car communication. In: 2007 IEEE 65th Vehicular Technology Conference - VTC2007-Spring, 22–25 April 2007, pp. 21–25 (2007)
20. Ren, G., Huang, Z., Cheng, Y., Zhao, X., Zhang, Y.: An integrated model for evacuation routing and traffic signal optimization with background demand uncertainty. J. Adv. Transp. **47**(1), 4–27 (2013)
21. Zhong, R.X., Sumalee, A., Pan, T.L., Lam, W.H.K.: Stochastic cell transmission model for traffic network with demand and supply uncertainties. Transportmetrica A Transp. Sci. **9**(7), 567–602 (2013)
22. Gipps, P.G.: A behavioural car-following model for computer simulation. Transp. Res. Part B Methodol. **15**(2), 105–111 (1981)
23. Xu, Y., Zheng, Y., Yang, Y.: On the movement simulations of electric vehicles: a behavioral model-based approach. Appl. Energy **283**, 116356 (2021). https://doi.org/10.1016/j.apenergy.2020.116356
24. Meszaros, F., Torok, A.: Theoretical investigation of emission and delay based intersection controlling and synchronising in Budapest. Period. Polytech. Transp. Eng. **42**(1), 37–42 (2014)
25. Jimenez, J.L., McClintock, P., McRae, G., Nelson, D.D., Zahniser, M.S.: Vehicle specific power: a useful parameter for remote sensing and emission studies. In: Ninth CRC On-Road Vehicle Emissions Workshop, San Diego, CA (1999)
26. Wyatt, D.W., Li, H., Tate, J.: Examining the influence of road grade on vehicle specific power (VSP) and carbon dioxide (CO_2) emission over a real-world driving cycle. No. 2013-01-1518. SAE Technical Paper (2013)
27. Zhang, L., Zeng, Z., Gao, K.: A bi-level optimization framework for charging station design problem considering heterogeneous charging modes. J. Intell. Connect. Veh. **5**(1), 8–16 (2022). https://doi.org/10.1108/JICV-07-2021-0009
28. Xu, Y., Ye, Z., Wang, C.: Modeling commercial vehicle drivers' acceptance of advanced driving assistance system (ADAS). J. Intell. Connect. Veh. **4**(3), 125–135 (2021). https://doi.org/10.1108/JICV-07-2021-0011
29. Lu, C., Liu, C.: Ecological control strategy for cooperative autonomous vehicle in mixed traffic considering linear stability. J. Intell. Connect. Veh. **4**(3), 115–124 (2021). https://doi.org/10.1108/JICV-08-2021-0012

Check for updates

Investigating Contributing Factors of Hard-Braking Events on Urban Road Network

Yue Zhou[1], Haiyue Liu[1], and Chuanyun Fu[2](✉)

[1] School of Transportation and Logistics, Southwest Jiaotong University, Chengdu, China
[2] School of Transportation Science and Engineering, Harbin Institute of Technology, Harbin, China
fuchuanyun@hit.edu.cn

Abstract. Hard-braking constitutes a critical surrogate measure of traffic safety on urban road networks. Efforts aiming to unveil the effects of contributing factors on the occurrence of hard-braking are inadequate. This study extracted the hard-braking event (HBE) and ordinary-braking event (OBE) by GPS trajectories from float cars. The effect of several factors on HBE was examined, including the factors of time, pre-braking behaviors, and road characteristics. The possibility of HBE was compared with that of OBE through binary logistic models (BLM). To further disclose the influence of factors, we considered the interaction between variables (BLM-VI) in modeling. The analysis results indicate that the BLM-VI is superior to the classical BLM in goodness-of-fit and factor interpretation. For the factors, peak hours on weekdays and daytime on weekends are positively linked to HBE, while driving at night is not. HBEs can be triggered by pre-braking behaviors such as speeding and approaching an intersection, but it is not likely to occur after changing lanes. Roads with work zones or intensive accesses can decrease the possibility of HBE. The factors of on-road parking, median divider, and one-way control have mixed effects on HBE when they interact with the factor of speed limits.

Keywords: Hard-braking event · Contributing factors · Urban road networks · Binary logistic model · Variable interaction

1 Introduction

Brake is a common reaction for drivers to obey the traffic controls or avoid incidents on-road. Among the braking behaviors, those whose decelerative velocity is extremely large are called hard-braking/harsh-braking. Hard-braking can indicate drivers strive to dodge a critical event involved or minimize its damage if it is inevitable [1]. As such, hard-braking typically serves as a kinematic indicator or surrogate measure of incidents in the urban area, such as conflict, crash, etc. [2, 3], and is widely used by the insurance industry [4].

Hard-braking tends to be triggered by both subjective and objective reasons. In some cases, hard-braking can be related to the lack of alert. For instance, past studies

Y. Bie et al. (Eds.): KES-STS 2022, SIST 304, pp. 112–122, 2022.
https://doi.org/10.1007/978-981-19-2813-0_12

revealed that hard-braking is associated with distracted driving [5], driving at high speed [1], and miscalculating the speed of the following vehicle [6]. For other cases, hard-braking is likely to be found in scenarios with safety risks that are inclined to be ignored. These scenarios contain driving on downhill roads, approaching an intersection, and dazzling due to sunlight in the daytime [7–9]. Due to these reasons, drivers usually have difficulties slowing down in advance of critical events because the response time-to-brake is increased [10].

Recent studies highlighted the effects of road characteristics on hard-braking. They suggested hard-braking is likely to be found on longer roads, secondary roads, and roundabouts with wide entries [2, 9]. However, their conclusions are limited in two aspects. First, nearly all of them focus on hard-braking frequency, while they failed to use an exposure (such as traffic flow) or adopt a categorical response to evaluate the actual hard-braking risk across scenarios. Second, considering the complexity of road networks, there are still factors of many kinds that need to be examined. Fortunately, naturalistic driving data have been used to extract braking behaviors [1, 7, 8], which would be inspired to fill the gaps above.

According to the review, we found several knowledge gaps among the analyses of hard-braking. Hereby, this study is conducted to explore the effect of factors including time, vehicle maneuvers prior to the brake (pre-braking behaviors), and road characteristics on hard-braking in urban road networks. The current study justifies the validity of using a large amount of float car data to extract braking events. Moreover, this study finds several pre-braking behaviors and road characteristics that are significantly associated with hard-braking.

2 Materials and Methods

2.1 Process of Raw Data

In the current study, floating car data (FCD) were used. The FCD were recorded through the vehicular GPS devices installed in taxis [11]. These data include the GPS trajectories of 5,755 taxis with a frequency of 0.1 Hz, which were generated in November 2016 in Chengdu. We restricted the study area to a rectangular zone because the road and environmental characteristics had been previously collected (Fig. 1). Each GPS point was matched to its nearest road by a map-matching algorithm. The specific steps of the algorithm are listed as follows [12, 13].

1) Prepare the digital Geo-map within the study area, the map layers consist of boundary, road geometry, facilities, etc.;
2) Manually measure the coordinate of the centerline and the center point for each road and intersection, respectively;
3) Choose a floating car and extract its trajectories from the first GPS point recorded through the timestamp, then load these GPS points to the Geo-map;
4) Match the GPS points to their nearest road centerlines with an improved Euclidean distance;
5) Check the trend of driving angle of the adjacent GPS points to filter U-turn and wrong-way driving behaviors, then remove these GPS points.
6) Redo step 1) to step 6) with another floating car.

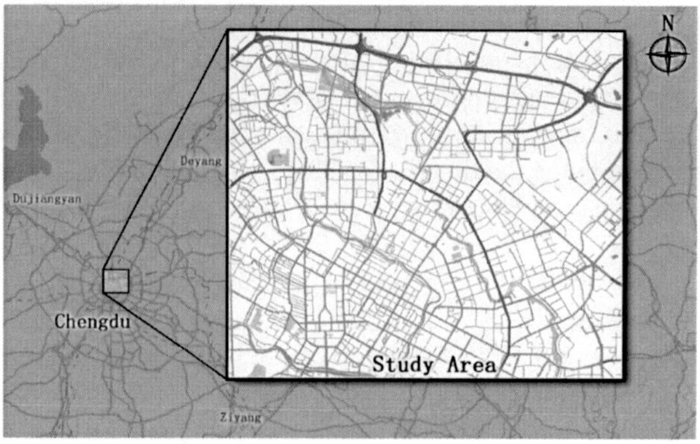

Fig. 1. Study area

2.2 Identification of Hard Braking Event

The braking events and hard-braking events are extracted from the FCD in the first two weeks of November (14 days). In order to minimize the possible error of identification, we deleted the FCD on curve segments. The identification steps of a braking event and hard braking event are given by:

1) First, select a pairwise point of adjacent GPS points generated by one vehicle. Note that the time interval of each pairwise of points is 10 s, which is determined by the frequency of the GPS device.
2) Record the instant speed of the start-point and the end-point of this pairwise point, respectively.
3) We assume that the vehicle movement between the two points is approximately close to uniformly accelerated motion so the accelerated velocity can therefore be calculated as

$$a_{m,m+1} = \frac{v_{m+1} - v_m}{t_{m,m+1}} \quad (1)$$

where $a_{m,m+1}$ is the accelerated velocity of the pairwise points $(m,\ m+1)$, v_{m+1} and v_m are the speed on point m (start-point) and point $m+1$ (end-point), respectively, $t_{m,m+1}$ is the difference of time between the two points.

4) The pairwise points whose accelerated velocity is <-0.2 g (g means the gravitational acceleration, i.e., $9.8\,\mathrm{m}/\mathrm{s}^2$) are defined as a hard braking event (HBE), while pairwise points whose accelerated velocity is [−0.1 g, −0.2 g] are treated as an ordinary-braking event (OBE). Both HBE and OBE are sorted as braking events. Pairwise points whose accelerated velocity is (−0.1 g, 0 g) are classified as slight slowdown rather than braking events.

The definition of HBE is achieved through the threshold of the accelerated velocity. However, such a threshold varies among the existing studies. Most studies defined the HBE with the threshold of [−0.75 g, −0.2 g], with the minimum acceptance of −0.2 g [14–16]. Botzer et al. [17] suggest a varying threshold of [−0.6 g, −0.2 g] due to the uncertainty of driving conditions. Here, we adopted a tolerable threshold of −0.2 g suggested by Olson et al. [16]. This is because the severe HBEs are quite rare in the current study area where is not associated with obvious changes in gradient.

Additionally, while the assumption of treating vehicle movement as uniformly accelerated motion may be problematic in reflecting the real-time path, it does not hamper the effectiveness of capturing HBE. Figure 2 presents the speed changes either set by our assumption or reflected by the actual movement. We provide several speed changes reflected by the possible actual paths (dash lines) for the study vehicle during the interval of the GPS record (10 s). For the threshold of −0.2 g that equals to the slope of the bold red line (assumption), no matter what the actual movement of the vehicle is, there will be at least one interval of the actual paths that has an accelerated velocity <-0.2 g (between the start-point and the end-point). In a sense, the assumption of uniformly accelerated motion can effectively capture the HBE during the interval of pairwise points, although not all the HBEs of a vehicle are fully recorded.

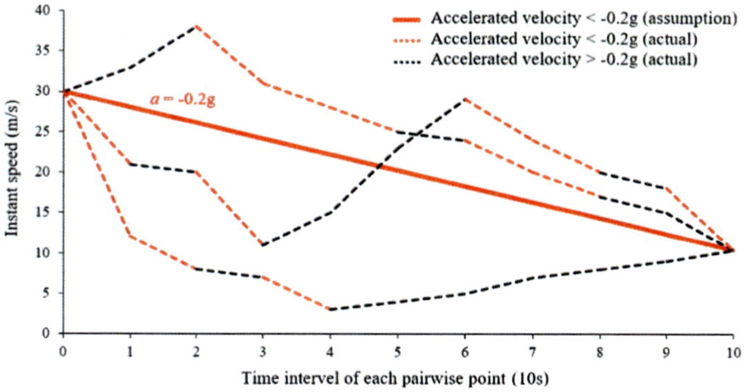

Fig. 2. The changes on vehicle speeds during the interval of the GPS record

The process of identification above is conducted to each float car through the plate number. Finally, the study recorded 12,424 HBEs and 344,823 OBEs extracted from the float cars during the study period.

2.3 The Measure of Potential Factors

Factors of time, driving behavior, traffic conditions, and road characteristics are associated with risky events such as crashes, speeding, conflicts, etc. [18–20], thus are considered in this study. All factors are processed to be categorical variables. Table 1 lists the description of these key factors.

2.4 Statistical Methods

This study aims to examine the effect of factors on triggering HBEs compared with OBE. To this end, we used a binary logistic model (BLM) to measure the effect of factors on HBE risk. In the binary logistic model, whether a braking event is HBE was selected as dependent variables. Besides, we also examined the interaction between categorical independent variables through a binary logit model with variable interaction (BLM-VI) to prevent any biased estimation. Note that the BLM and BLM-VI only include the variables with variance inflation factor (VIF) <5.

Table 1. Description and explanation of key factors

Type	Factor	Explanation
Time	Time of day	Whether the HBE.OBE occurs on daytime (=0), peak hours (=1), or at night (=2)
	Weekend	Whether the HBE/OBE occurs on weekend (=0 if no and = 1 if yes)
Pre-braking behavior	Speeding	Speeding before an HBE/OBE (=0 if no and = 1 if yes)
	Lane-changing	Chang lanes or steer before an HBE/OBE (=0 if no and = 1 if yes)
	Approaching an intersection	The distance between an HBE/OBE and an intersection is less than 30 m (=0 if no and = 1 if yes)
Road characteristics	Speed limit	The posted speed limit of the road where an HBE/OBE occurs (=0 if ≥ 60 km/h, = 1 if 30–60 km/h, = 2 if ≤ 30 km/h)
	Work zone	HBE/OBE is located nearby a work zone (=0 if no and = 1 if yes)
	One-way control	HBE/OBE occurs on a road with one-way control (=0 if no and = 1 if yes)
	On-road parking	HBE/OBE occurs on a road where on-road parking is observed (=0 if no and = 1 if yes)
	Median divider	HBE/OBE occurs on a road with median divider (=0 if no and = 1 if yes)
	Accesses per km	The number of accesses per each kilometer on the road where an HBE/OBE occurs (=0 if ≤ 10, = 1 if 10–13.5, = 2 if ≥ 17)

3 Results and Discussions

3.1 Estimates of Contributing Factors of HBE

Table 2 lists the estimates and goodness-of-fit of BLM and BLM-VI. The factor of speeding is eliminated from BLM since it is severely multicollinear with the factor of speed limits. The results show that driving at night ($\beta = -0.643$), implementing lane-changing maneuver ($\beta = -0.716$), roads with 10–13.5 accesses per km ($\beta = -0.111$) and $\geq$17 accesses per km ($\beta = -0.215$), and work zone ($\beta = -0.141$) significantly decrease the possibility of HBE. Conversely, approaching an intersection ($\beta = 0.292$), speed limits of 30–60 km/h ($\beta = 0.406$) and $\leq$30 km/h ($\beta = 1.231$), on-road parking ($\beta = 0.137$), and median divider ($\beta = 0.238$) significantly increase the possibility of HBE. It is also noticed that driving on-peak hours, roads with 13.5–17 accesses per km, and weekends are not significantly related to the occurrence of HBE.

It is noted that four factors that do not interact with others in both models show a similar trend in affecting HBEs. However, the remaining factors in BLM-VI are interactive and show discrepant signs and significance as compared with those in BLM. More specifically, estimates of BLM-VI indicate that peak hours on weekdays ($\beta = 0.091$) are positively associated with HBEs, while the effect of peak hours in BLM is insignificant overall. Speeding is removed in BLM due to multicollinearity. Conversely, it could be a good indicator in BLM-VI if this behavior occurs on roads with a speed limit of $\leq$30 km/h ($\beta = 1.039$) or $\geq$60 km/h ($\beta = 1.069$). On-road parking and median divider are positively associated with HBEs in BLM. However, on-road parking is only positively linked to HBEs on roads whose speed limit is <30 km/h ($\beta = 0.367$) or 30–60 km/h ($\beta = 0.179$), while median divider merely increases HBE risk on $\geq$60 km/h roads ($\beta = 0.337$). We also notice that one-way control, which is insignificant in BLM, has varied effects across the speed limits as estimated by BLM-VI ($\beta = -1.046$ for speed limits of <60 km/h, $\beta = -0.395$ for speed limits of 30–60 km/h and $\beta = 0.418$ otherwise). In addition to the difference of estimates, it shows that BLM-VI is superior to BLM in goodness-of-fit (7918 vs. 10,431 for AIC; −3,933 vs. −5,203 for log-likelihood). This justifies the necessity and benefit of accounting for variable interaction. As a result, we adopt the estimates of BLM-VI to interpret the contribution of factors.

Table 2. Estimation results of BLM and BLM-VI.

BLM			BLM-VI		
Variable	Mean (β)	S.D.	Variable	Mean (β)	S.D.
Intercept	−3.091*	0.025	Intercept	−3.176*	0.027
Time of day			Time of day × Weekend		
[peak hours]	0.037	0.025	[night] × [yes]	−0.573*	0.036
[night]	−0.643*	0.023	[peak hours] × [yes]	−0.017	0.044
[daytime]	0[a]	–	[daytime] × [yes]	0.065*	0.026

(continued)

Table 2. (*continued*)

BLM			BLM-VI		
Variable	Mean (β)	S.D.	Variable	Mean (β)	S.D.
–	–	–	[night] × [no]	−0.638*	0.028
–	–	–	[peak hours] × [no]	0.091*	0.03
–	–	–	[daytime] × [no]	0[a]	–
Lane-changing			Lane-changing		
[yes]	−0.716*	0.02	[yes]	−0.677*	0.02
[no]	0[a]	–	[no]	0[a]	–
Speed limit			Speeding × Speed limit		
[30–60 km/h]	0.406*	0.029	[yes] × [≤30 km/h]	1.039*	0.048
[≤ 30 km/h]	1.231*	0.304	[yes] × [30–60 km/h]	−0.075	0.049
[≥ 60 km/h]	0[a]	–	[yes] × [≥60 km/h]	1.069*	0.038
–	–	–	[no] × [≤30 km/h]	2.691*	0.066
–	–	–	[no] × [30–60 km/h]	1.516*	0.045
–	–	–	[no] × [≥60 km/h]	0[a]	–
Approaching an intersection			Approaching an intersection		
[yes]	0.292*	0.024	[yes]	0.295*	0.024
[no]	0[a]	–	[no]	0[a]	–
On-road parking			On-road parking × Speed limit		
[yes]	0.137*	0.03	[yes] × [≤30 km/h]	0.367*	0.055
[no]	0[a]	–	[yes] × [30–60 km/h]	0.179*	0.065
–	–	–	[yes] × [≥60 km/h]	−0.375*	0.055
–	–	–	[no] × [≤30 km/h]	0[a]	–
–	–	–	[no] × [30–60 km/h]	0[a]	–
–	–	–	[no] × [≥60 km/h]	0[a]	–
Median divider			Median divider × Speed limit		
[yes]	0.238*	0.02	[yes] × [≤30 km/h]	−0.297*	0.089
[no]	0[a]	–	[yes] × [30–60 km/h]	−0.205*	0.078
–	–	–	[yes] × [≥60 km/h]	0.337*	0.022
–	–	–	[no] × [≤30 km/h]	0[a]	–
–	–	–	[no] × [30–60 km/h]	0[a]	–
–	–	–	[no] × [≥60 km/h]	0[a]	–
NA			One-way control × Speed limit		
–	–	–	[yes] × [≤30 km/h]	−1.046*	0.187
–	–	–	[yes] × [30–60 km/h]	−0.395*	0.106

(*continued*)

Table 2. *(continued)*

BLM			BLM-VI		
Variable	Mean (β)	S.D.	Variable	Mean (β)	S.D.
–	–	–	[yes] × [≥60 km/h]	0.418*	0.069
–	–	–	[no] × [≤30 km/h]	0[a]	–
–	–	–	[no] × [30–60 km/h]	0[a]	–
–	–	–	[no] × [≥60 km/h]	0[a]	–
Accesses per km			Accesses per km		
[10–13.5]	−0.111*	0.026	[10–13.5]	−0.261*	0.028
[13.5–17]	0.029	0.027	[13.5–17]	−0.04	0.028
[≥17]	−0.215*	0.027	[≥17]	−0.167*	0.026
[≤10]	0[a]	–	[≤10]	0[a]	–
Work zone			Work zone		
[yes]	−0.141*	0.04	[yes]	−0.082*	0.039
[no]	0[a]	–	[no]	0[a]	–
Log-likelihood	−5,203		−3,933		
AIC	10,431		7,918		

Note: * indicates the variable is significant at 95% confidence interval; [a] means the factor is set as reference so its coefficient equals to 0; NA means the variable is not available.

3.2 Explanation of Contributing Factors

For the effect of time-related factors, we found that peak hours on weekdays and daytime on weekends are associated with HBEs. The two periods usually have more traffic activities and larger traffic volumes, which could cause a higher risk of traffic incidents [21]. Conversely, driving at night tends to encounter fewer incidents and experience better traffic conditions, consequently reducing the HBE risk.

With respect to the pre-braking behaviors, lane-changing maneuver requires the vehicle to move slower and carefully steer to the target lane, which is not likely to encounter a sudden brake. Speeding, as expected, can trigger HBEs. This is attributed to the drivers could have difficulties responding fast while speeding and need to implement emergency brakes if an unexpected incident occurs [22]. Note that speeding on roads with a 30–60 km/h speed limit is not linked to HBEs, which may be that fewer incidents are found on these roads. Drivers who approach an intersection are more likely to encounter HBEs, possibly owing to the situation that drivers are unaware of the forthcoming change on the signal phase when they are about the get across the intersection [23].

Interestingly, speed limits interact with several road characteristics and yield mixed effects on HBEs. A reasonable explanation is that these characteristics have varied functions across the road class. Therefore, for the factor of on-road parking, moving vehicles are more easily affected by vehicles entering/leaving a parking space set on secondary or branch roads that have fewer available lanes, increasing HBE risk on roads with lower speed limits as well. Median divider and one-way control could raise the travel speed so they may be more likely to cause HBEs on arterial roads where rear-end conflicts are common [24]. The findings also show that access density and work zone are associated with lower HBE risk, which implies that the factors can decrease the vehicle speed and eventually hamper the occurrence of HBE.

4 Conclusion

Hard-braking events are usually related to safety incidents thus serve as surrogate measures. Our study investigated the effect of factors including time, road design, and pre-braking behavior on HBEs in urban areas. The effect of contributing factors is explored by binary logistic models with variable interaction.

Results demonstrate that HBEs tend to occur at the daytime on weekends and peak hours on weekdays, while is less likely found at night. Speeding on roads with a speed limit of $\leq$30 km/h or $\geq$60 km/h and approaching an intersection can trigger HBEs, yet changing lanes prior to the brake does not lead to an HBE. On-road parking tend to cause HBEs on roads with lower speed limits ($<$60 km/h); Median divider and one-way control are associated with the occurrence of HBE if the corresponding roads have a speed limit of $\geq$60 km/h. Also, HBE risk reduces as the roads are observed with intensive accesses or work zone. We also highlight the consideration of variable interaction in modeling harsh driving behaviors to obtain a robust estimation.

There are two limitations that need to be further revised. First, the modeling approach can be extended with random effects/parameters approaches which are able to capture unobserved heterogeneity [25–27]. Second, the driver's socio-demographic information, enforcement, and underlying spatial-temporal effects can be combined to extend the modeling approach [28–30]. Also, we recommended that countermeasures such as warning system can be used to restrict harsh driving behaviors [31].

Acknowledgements. This study was jointly sponsored by the National Natural Science Foundation of China (Grant No. 71801182) and the Fundamental Research Funds for the Central Universities (FRFCU5710000111).

References

1. Knotaxi, A., Ziakopoulos, A., Yannis, G.: Trip characteristics impact on the frequency of harsh events recorded via smartphone sensors. IATSS Res. **45**, 574–583 (2021)
2. Kamla, J., Parry, T., Dawson, A.: Analysing truck harsh braking incidents to study roundabout accident risk. Accid. Anal. Prev. **122**, 365–377 (2019)
3. Fu, C., Sayed, T., Zheng, L.: Multivariate Bayesian hierarchical modeling of the non-stationary traffic conflict extremes for crash estimation. Anal. Methods Accid. Res. **28**, 100135 (2020)
4. Boquete, L., Rodríguez-Ascariz, J.M., Barea, R., Cantos, J., Miguel-Jiménez, J.M., Ortega, S.: Data acquisition, analysis and transmission platform for a pay-as-you-drive system. Sensors **10**(6), 5395–5408 (2010)
5. Haque, M.M., Oviedo-Trespalacios, O., Debnath, A.K., Washington, S.: Gap acceptance behavior of mobile phone-distracted drivers at roundabouts. Transp. Res. Rec. **2602**(1), 43–51 (2016)

6. Maslać, M., Antić, B., Lipovac, K., Pešić, D., Milutinović, N.: Behaviours of drivers in Serbia: non-professional versus professional drivers. Anal. Methods Accid. Res. **52**, 101–111 (2018)
7. Mase, J.M., Majid, S., Mesgarpour, M., Torres, M.T., Figueredo, G.P., Chapman, P.: Evaluating the impact of heavy goods vehicle driver monitoring and coaching to reduce risky behaviour. Accid. Anal. Prev. **146**, 105754 (2020)
8. Ziakopoulos, A.: Spatial analysis of harsh driving behavior events in urban networks using high-resolution smartphone and geometric data. Accid. Anal. Prev. **157**, 106189 (2021)
9. Fu, C., Zhang, Y., Bie, Y., Hu, L.: Comparative analysis of driver's brake perception-reaction time at signalized intersections with and without countdown timer using parametric duration models. Accid. Anal. Prev. **95**, 448–460 (2016)
10. Fu, C., Sayed, T.: Comparison of threshold determination methods for the deceleration rate to avoid a crash (DRAC)-based crash estimation. Accid. Anal. Prev. **153**, 106051 (2021)
11. Fu, C., Liu, H.: Investigating distance halo effect of fixed automated speed camera based on taxi GPS trajectory data. J. Traffic Transp. Eng. (English Edition) (2021, forthcoming)
12. Zhou, Y., Jiang, X., Fu, C., Liu, H.: Operational factor analysis of the aggressive taxi speeders using random parameters Bayesian LASSO modeling approach. Accid. Anal. Prev. **157**, 106183 (2021)
13. Fu, C., Zhou, Y., Xu, C., Cui, H.: Spatial analysis of taxi speeding event using GPS trajectory data. In: Proceedings of the 2019 IEEE Intelligent Transportation Systems Conference, pp. 122–127. IEEE, New York (2019)
14. Grygier, P.A., et al.: Study of heavy truck air disc brake effectiveness on the national advanced driving simulator. In: Proceedings of the 20th ESV Conference (2007)
15. Xu, Y., Ye, Z., Wang, C.: Modeling commercial vehicle drivers' acceptance of advanced driving assistance system (ADAS). J. Intell. Connect. Veh. **4**(3), 125–135 (2021). https://doi.org/10.1108/JICV-07-2021-0011
16. Pan, C., Xu, J., Fu, J.: Effect of gender and personality characteristics on the speed tendency based on advanced driving assistance system (ADAS) evaluation. J. Intell. Connect. Veh. **4**(1), 28–37 (2021). https://doi.org/10.1108/JICV-04-2020-0003
17. Botzer, A., Musicant, O., Mama, Y.: Relationship between hazard-perception-test scores and proportion of hard-braking events during on-road driving – an investigation using a range of thresholds for hard-braking. Accid. Anal. Prev. **132**, 105267 (2019)
18. Zhou, Y., Fu, C., Jiang, X., Mao, C., Liu, H.: Road factor analysis of taxi speeding behavior considering spatial effect. China Saf. Sci. J. **31**(3), 162–170 (2021)
19. Fu, C., Tarek, S.: Bayesian dynamic extreme value modeling for conflict-based real-time safety analysis. Anal. Methods Accid. Res. **34**, 100204 (2022)
20. Fu, C., Tarek, S., Zheng, L.: Multi-type Bayesian hierarchical modeling of traffic conflict extremes for crash estimation. Accid. Anal. Prev. **160**, 106309 (2021)
21. Høye, A.K., Hesjevoll, I.S.: Traffic volume and crashes and how crash and road characteristics affect their relationship – a meta-analysis. Accid. Anal. Prev. **145**, 105668 (2020)
22. Chevalier, A., et al.: Exploration of older drivers' speeding behaviour. Anal. Methods Accid. Res. **42**(3), 532–543 (2016)
23. Xiang, W., Yan, X., Weng, J., Li, X.: Effect of auditory in-vehicle warning information on drivers' brake response time to red-light running vehicles during collision avoidance. Anal. Methods Accid. Res. **40**, 56–67 (2016)
24. Liu, H., Fu, C., Jiang, C., Zhou, Y., Mao, C., Zhang, J.: Bayesian hierarchical spatial count modeling of taxi speeding events based on GPS trajectory data. PLoS ONE **15**(11), 1–17 (2020)
25. Fu, C., Sayed, T.: Random parameters Bayesian hierarchical modeling of traffic conflict extremes for crash estimation. Accid. Anal. Prev. **157**, 106159 (2021)

26. Fu, C., Sayed, T.: Multivariate Bayesian hierarchical Gaussian copula modeling of the non-stationary traffic conflict extremes for crash estimation. Anal. Methods Accid. Res. **29**, 100154 (2021)
27. Ali, Y., Zheng, Z., Haque, M.M.: Modelling lane-changing execution behaviour in a connected environment: a grouped random parameters with heterogeneity-in-means approach. Commun. Transp. Res. **1**, 100009 (2021). https://doi.org/10.1016/j.commtr.2021.100009
28. Yu, B., Chen, Y., Bao, S.: Quantifying visual road environment to establish a speeding prediction model: an examination using naturalistic driving data. Accid. Anal. Prev. **129**, 289–298 (2019)
29. Jiang, X., Liu, H., Zhou, Y., Fu, C.: Systematic review on anti-speeding countermeasure research. China J. Highw. Transp. **33**(3), 1–31 (2020)
30. Ayele A.E., Jiang, X., Fu, C.: Spatial point pattern analysis of traffic violations in Luzhou City, China. Transp. Lett., 1–10 (2021)
31. Zhao, X., Li, X., Chen, Y., Li, H., Ding, Y.: Evaluation of fog warning system on driving under heavy fog condition based on driving simulator. J. Intell. Connect. Veh. **4**(2), 41–51 (2021). https://doi.org/10.1108/JICV-11-2020-0012

Usage Pattern Analysis of e-scooter Sharing System: A Case Study in Gothenburg, Sweden

Gentrina Peci[1], Sadia Ali[1], Jieyu Fan[2](✉), and Jie Zhu[1]

[1] Department of Architecture and Civil Engineering, Chalmers University of Technology, 412 96 Gothenburg, Sweden
[2] Faculty of Engineering, Computer Science and Psychology, Department of Human Factors, Ulm University, 89069 Ulm, Germany
fan.jieyu@uni-ulm.de

Abstract. Swedish cities are embracing shared micro-mobility systems (SMMS) such as e-scooters sharing systems to promote sustainable travel behavior in urban contexts with corresponding infrastructure planning. SMMS is associated with various social, environmental, and economic benefits, as well as providing solutions for the first- and the last-mile problem of using public transit. This study analyzes the usage patterns of e-scooter systems, based on the scooter operation data of VOI company in Gothenburg, Sweden. The used data cover the transaction data of two and half months during the summer and include over five hundred thousand valid trip records. The result shows that most trips travel a distance between 0.5–1.8 km while the duration lasts between 4–7 min. Fridays and Saturdays are the most popular days while Sunday is the least popular day. The number of trips on Sundays decreases by about 60% compared to Fridays and Saturdays. Moreover, the e-scooters are used to varying degrees in the different areas of Gothenburg. The e-scooters are used at a much higher extent in central Gothenburg compared to areas outside the city center. This can be due to several different factors such as location, land use, and accessibility. Lastly, the results show that the e-scooters are not primarily used for commuting but rather for leisure, which can be seen in the average distance and duration of the entire zone as well as the temporal distribution.

Keywords: Micro-mobility · Shared micro-mobility · E-scooters · Usage patterns

1 Introduction

Every year more than one hundred million people are affected by climate catastrophes. According to the Red Cross [1], up to two hundred million people may need emergency aid by 2050 if the world does not act now. In connection with the climate changes that are currently taking place because of global warming, multiple countries around the world have decided to reduce their emissions. Sweden has a goal of attaining net-zero emissions of GHG into the atmosphere by 2045 [2–4]. To achieve this, Sweden must reduce their emissions from other sectors including the transport sector, which accounts

Y. Bie et al. (Eds.): KES-STS 2022, SIST 304, pp. 123–132, 2022.
https://doi.org/10.1007/978-981-19-2813-0_13

for one-third of all greenhouse gases emitted [5]. According to a study conducted by traffic analysis [6], the majority of all journeys that take place in Sweden are carried out by car. This corresponds to 55% of all trips made in 2020. The remaining 45% were distributed rather evenly between public transport, walking, and cycling, see Fig. 1.

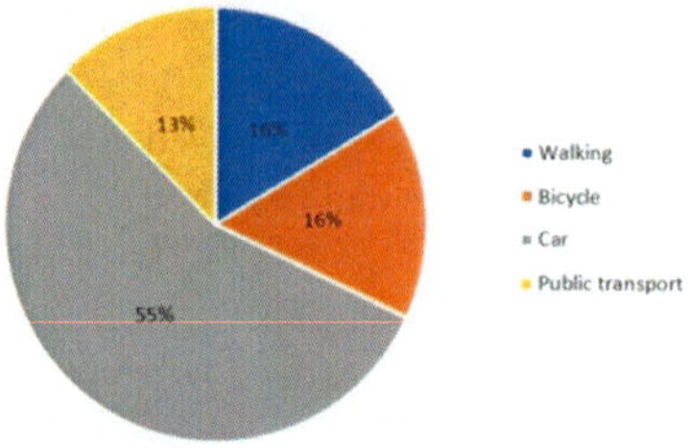

Fig. 1. Travel mode split structure in Sweden 2020.

There are several different approaches available to reduce the utilization of cars which has a negative effect on the environment. One of them is the use of shared micro-mobility systems such as electric scooter sharing mobility. According to Mansky [7, 8], electric scooters have existed for over one hundred years, but it is not until recent years that the demand for these micro-mobility devices has increased radically. The modern-day electric scooters were first introduced in the United States in 2017 and have since spread rapidly around the world [9, 10]. Electric scooters reached the Swedish market shortly after when the Swedish company VOI launched their own electric scooters in 2018. Without doubts, shared e-scooter systems (SESS) can provide convenient travel tools for users to reduce travel time or cost in many situations, as a result of its flexibility. The convenience and flexibility of shared electric scooter systems bring prosperity by implementing them in urban contexts, especially in the European Union, EU. The aim of this thesis is to answer the above-mentioned research questions based on massive real usage data of SESS. We got access to the transaction data of e-scooter systems in Gothenburg from the open API of the VOI company. Based on the unique data, we mainly investigate the spatiotemporal usage patterns of SESS in Gothenburg. Particularly, this study reveals the potential differences in the usage patterns in different urban areas with different built environments.

2 Literature Review

As urban areas throughout the world continue to expand rapidly, and the demand for the existing transportation network is increasing, micro-mobility is gaining attention as a practical alternative. Micro-mobility is defined as a short-distance transport and can be described as "Personal transportation using devices and vehicles weighing up to 350 kg and whose power supply, if any, is gradually reduced and cut off at a given speed limit which is no higher than 45 km/h" [11]. In micro-mobility, human-powered vehicles are used exclusively, such as bikes, e-scooters, skateboards, e-rickshaws, etc. Micro mobility is offering attractive solutions for "first and last-mile connectivity," as well as assisting in the reduction of traffic [12].

Moreover, the term "micro-mobility" gained popularity around the year 2016 when Dediu [10] an American analyst, connected the term to sharing vehicles such as bicycles and scooters. The term "micro," according to Dediu [10] refers to either the short distances that are traveled by such vehicles or the vehicle itself. In recent years, the usage of micro vehicles has increased drastically in connection with the introduction of electric scooters. Many micro vehicles are owned by private individuals who utilize them for their daily trips. In addition to the privately owned vehicles, it is also common to find electric scooters and bicycles on the streets in the city that are available to rent [11]. Shared micro-mobility system (SMMS), is the system of sharing using low-speed modes such as electric scooters (e-scooters) and electric bikes (e-bikes). This innovative transportation system provides users temporary access to small modes of transportation to satisfy short-distance travel demands [12, 13] and thus has an impact on users' travel behavior and decision-making. Shared bike-sharing can be divided into station-based (docked), dockless, and hybrid bike-sharing systems. Scooter sharing, on the other hand, can be divided into standing electric scooter sharing and moped-style scooter sharing. The main difference between docked -and dockless sharing systems is that docked sharing system provides one-way station-based service, while dockless sharing system enables checking out the sharing mode and returning in any locations within a predefined geographic region [14].

Furthermore, Bielinski and Wazna [15] concluded that the availability of shared micro-mobility needs to be considered and improved to reach out to as many social groups as possible. Urban areas are increasingly affected by different issues such as traffic congestion, car accidents, space occupied by cars, air pollution and external transportation costs. Accordingly, Bielinski and Wazna [15] state that increasing the availability of these services will support municipal administrations in addressing the challenges associated with urbanized areas.

Additionally, Insurance Soved Blog [16] reports that 31% of the pedestrians in the Australian city Adelaide are uncomfortable when sharing the sidewalk with electric scooters, with a proportion increasing with the pedestrian's age. Moreover, the study shows that 29% of the pedestrians have been forced to move quickly aside to avoid colliding with electric scooters. The study also states that up to 40% of the e-scooters were not parked as per the instructions of the operator, which posed a potential safety risk to the pedestrians [16]. According to the Swedish Transport Agency [17], the number of accidents with electric scooters has risen to 1 000 accidents in the past year. In 2016–2018, approximately twenty personal injury accidents occurred per year where an electric scooter was involved. In the following years, the number of injury accidents increased, and 1 056 accidents involving electric scooters occurred from August 2021 [17].

3 Study Area and Data

The area examined in this study is the city Gothenburg located in the western part of Sweden. Gothenburg is the second-largest city, with an area of 448 km^2 and a population of 580,000 inhabitants [18]. Gothenburg is divided into 4 major urban areas consisting of Northeast, Centrum, Southwest, and Hisingen, as shown by Fig. 2.

Since the purpose of this thesis is to investigate the usage patterns of VOI's electric scooters in Gothenburg the study area will be limited to VOI's own zone which can be seen in Fig. 3. VOI has limited the use of their electric scooters to Gothenburg's inner city as well as some other adjacent areas. This area will henceforth be referred to as the "VOI zone". As presented in Fig. 3 the zone lies within major roads such as E20, E6 and E45. The zone also contains all the major districts that lie within Gothenburg's inner city. The zone has an area of 152.2 km^2 and a circumference of 71 km.

Fig. 2. Map of the study area.

Fig. 3. Land use in central Gothenburg [20].

The different zones in Gothenburg include the inner city and the middle city. The inner city is the most central part of Gothenburg and constitutes most of our study area, while the middle city is directly adjacent to the inner city. These areas consist mainly of mixed urban development that hold housing, workplaces, offices, trade, facilities, and various public services. In addition to buildings, mixed urban areas also include local streets, pedestrian, and bicycle paths, as well as public places such as squares and parks [20]. Furthermore, Fig. 3 illustrates that there are several important public transport hubs, marked as red dots, within the study area. Other areas of interest are parks and sports grounds, marked in green, as well as university campuses which are marked with many small red dots. Moreover, even though there are residentials/housing in all areas the proportion of households is much smaller in the city center compared to the rest of the zone as there is a greater focus on, for example, trade within the city center [21, 22].

3.1 Extracting Trip Transactions from Position Data

The raw data provided by VOI was downloaded from open API data in form of positioning data. Meaning it only contained the GPS information about the electric scooters that were not in use. Once an electric scooter was in use it was excluded from the list. Moreover, the data was downloaded at a frequency of 10–15 s, making it possible to record many trips to provide a fair representation of the situation. By downloading the data at such high frequency, it was ensured that the start and end coordinates of each trip were included in the dataset. It also made it possible to extract the distance based on the position data. Lastly, by knowing the download frequency the time stamps of each trip could also be extracted. In total, approximately two months' worth of positioning data was transformed into transaction data which was then used to perform the different analyses. The dataset contained 762 565 trips. Some trips were, however, abnormal, indicating the need of clearing the dataset from outliers. This was done by limiting the travel distance, duration, and speed. According to VOI's terms of service, users can only travel for 45 min continuously at a maximum speed of 20 km/h, thus setting the maximum duration to 45 min and limiting the speed to 20 km/h. The distance was also limited to a maximum of 10 km. Once all the outliers were cleared out of the transaction data, the number of trips were reduced from 762 565 to 532 938. The analysis was then performed using the new cleared dataset.

4 Results

4.1 Usage Demand and Trip Characteristics

The result is based on data that begins on the 14th of May and ends on the 20th of July. Figures 4 and 5 show the number of trips that have taken place in each zone. The dark red color in Fig. 4 indicates that many trips have taken place in that specific area while yellow indicates that the number of trips in that area is small and lastly orange is somewhere in between. The exact intervals can be seen in Fig. 4. From the figure and the diagram, it is clear that the areas with the largest number of trips are Olivedal-Haga, Inom Vallgraven and Johanneberg in descending order. On the other hand, the areas with the lowest number of trips are Skår, Delsjöområdet, and Tingstadsvassen-Hisingsbron in descending order. The other areas are somewhere in between. Nevertheless, a clear pattern is distinguished where the areas located in the city center have a high number of trips while the areas towards the edges have a much smaller number of trips. Furthermore, an exception to this trend is the area Kvillebäcken that has a larger number of trips despite being at the edge of the zone as well as being surrounded by zones that have a smaller number of trips.

Based on the approximately two months' worth of data, the distribution of trip distance and duration was calculated as well as the temporal distribution for all the

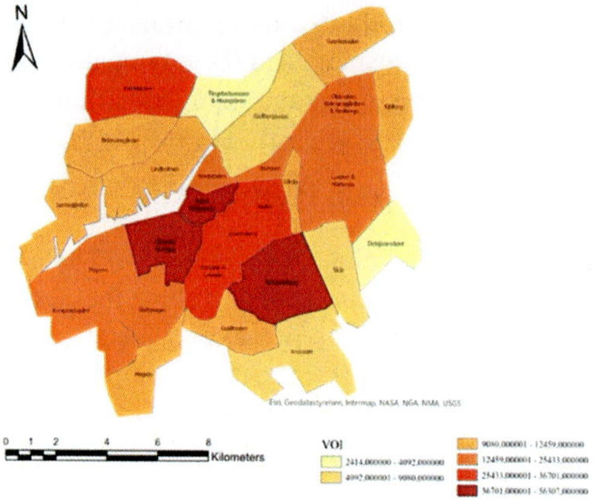

Fig. 4. Number of Voi trips for different zones.

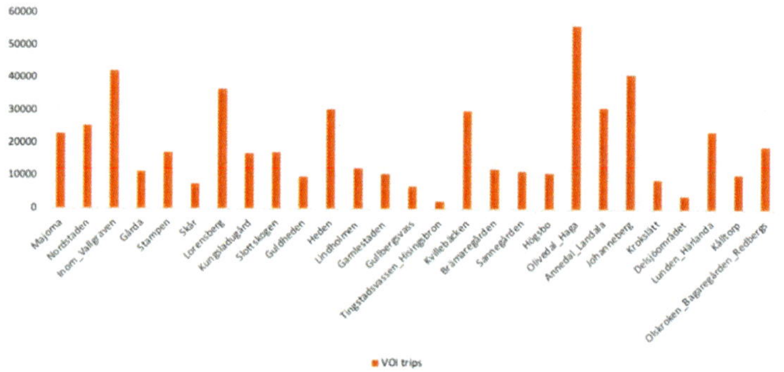

Fig. 5. Number of VOI trips in Gothenburg for different zones based on transaction data.

zones within the VOI zone. Figure 6 shows the result of each of those calculations. For the network distance, the results show that most trips travel a distance between 0.5–1.8 km, and the number of trips decreases gradually with the distance. Furthermore, most trips have a duration between 4–7 min. Similar to previous mentioned pattern, most trips are shorter rides, and the number of trips decreases as the duration increases. Lastly, the temporal distribution shows during what day of the week and what time of the day most of the users ride the e-scooters. Figure 6 (c) shows that the most popular days are Fridays and Saturdays where the number of trips is around 800, while the least popular day is Sundays where the number of trips decreases by about 60% compared to Fridays and Saturdays. Furthermore, the graph also shows that the number of trips increases during the day. The least number of trips occurs in the morning between 06.00–09.00 with about 200–300 trips. During noon, the number of trips has increased by about 150% to around 500 trips. The maximum peak hour occurs in the afternoon between 15.00–19.00 with around 600–800 trips where the number of trips has increased by 300–350% compared to the morning peak hour. Additionally, it can also be observed that the afternoon peak hour has a longer duration during the weekends (Friday and Saturday nights) and lasts until 22.00.

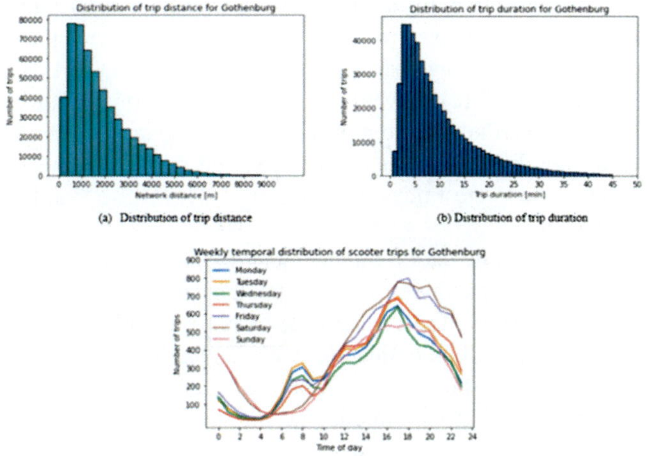

(a) Distribution of trip distance

(b) Distribution of trip duration

(c) Temporal distribution of scooter trips

Fig. 6. Spatial and temporal distributions for the entire Voi zone.

4.2 Differences of Using SESS Different Zones

Figure 7 shows that all areas have a trip distance distribution that is skewed to the right except Delsjöområdet. This areas distribution reminiscent of the right-skewed shape however because of the bar standing out to the left it deviates from the traditional shape. The right-skewed distribution indicates that the number of trips decreases as the distance increases.

A clear difference regarding the travel distances, as well as the number of trips, can be seen between the areas that have a high demand compared to the areas with a small demand (top three and bottom three). Areas with a high demand tend to travel very short distances between 0.8–1.5 km. For instance, in Fig. 7 (b), about 5000 of all trips that take place in Olivedal Haga are between 0.8–1.2 km. Similar occurrences can also be seen for Inom Vallgraven and Johanneberg, see Table 4. Trips that take place in the three low demand areas tend to have slightly longer distances ranging between 1–2.5 km apart from Delsjöområdet, see Fig. 7 (e). In this area, most trips are about 1.5 km, followed by a distance between 3–4 km. In addition, the number of trips in the low demand areas is significantly less than the number of trips that take place in high demand areas. Among the low demand areas, Tingstadsvassen illustrated in Fig. 7(d–f) has the highest number of trips occurring at distances between 0.8–1.5 km where the number of trips amounts to approximately 1400. For the other two areas, the number of trips amounts to approximately 500 respectively. This means that there are almost eight times more trips occurring in the three high demand areas compared to the three low demand areas. Moreover, VOI has limited the trips to a max distance of 10 km. However, the results show that the longest distance among all the areas was approximately 9.3 km which yet again occurred in Delsjöområdet.

Since all the distributions have a right skew, the average value is a little larger than the median value. By comparing the different values with the figures for both trip distance and trip duration, it may be considered that the median is more representative than the mean. For example, the average trip distance for Johanneberg is 1701 m while the median is 1391 m, and Fig. 7 (c) shows that there are more trips occurring at 1400 m compared to 1700 m. Nevertheless, the most correct parameter is the mode value of all trips. The modes for the different areas can be seen in Table 4, which shows that trips in Delsjöområdet have the longest travel distance of 1500 m while Skår has the shortest travel distance of 600 m.

Small values of standard deviation indicate that the dispersion of the data from the mean value is small, while high values indicate that the data has a great variance which obtains more uncertain results. The standard deviation for both the trip distance and duration is high for all areas, thus indicating that the data has a large dispersion. The area with the lowest value for standard deviation is Skår since most of the values are on the left side of the data, while Delsjöområdet has the highest standard deviation, which is expected since its data is very dispersed.

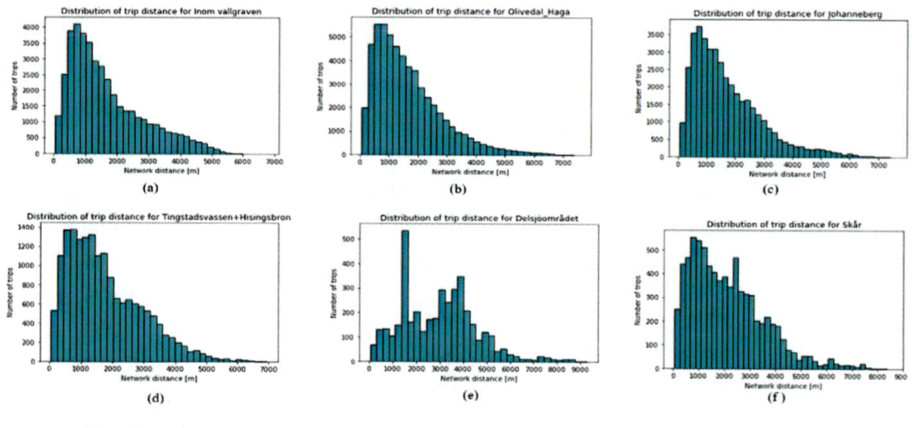

Fig. 7. Trip distance for the three areas with the biggest and smallest demand.

The percentiles show how the data is distributed within different intervals. In this study, it has been chosen to report for the 25th, 50th and 75th percentiles. For trip distance, it appears that 25% of all trips for the three high demand areas have a distance less than approximately 800 m, while 50% of all trips are shorter than approximately 2000 m. Lastly 75% of all trips are shorter than 2350 m. For the three low demand areas, there are slightly larger differences between the different areas. This is due to Delsjöområdet, which as previously mentioned, deviates from the traditional right-skewed distribution. However, in general, it can be said that about 25% of all trips are shorter than 1130 m while 50% of all trips are shorter than 2140 m and lastly 75% of all trips are shorter than 3140 m.

Table 4. Statistical parameters for trip distance.

Area	Nr of trips	Mean	Mode	Std	Min	25%	50%	75%	Max
Inom vallgraven	42 274	1676	800	1178	52	770	1343	2344	6778
Olivedal-Haga	56 307	1649	1000	1172	49	760	1374	2232	7568
Johanneberg	41 251	1701	800	1215	51	779	1391	2334	7614
Tingstadsvassen-Hisingsbron	17,330	1763	800	1187	49	830	1491	2511	6931
Delsjöområdet	4092	2978	1500	1597	54	1591	3066	3951	9289
Skår	7738	2107	600	1431	53	961	1864	2950	8349

5 Conclusion

Based on the transaction data of e-scooter sharing system in Gothenburg, this study investigates the usage patterns of the systems in both spatial and temporal dimensions. The main findings can be summarized as:

- The results show that e-scooters are used to varying degrees in the different areas of Gothenburg. E-scooters are used to a much higher extent in central Gothenburg compared to areas outside the city centre. This can be due to several different reasons such as location, land use, and accessibility.
- When e-scooters were first introduced, they were described as a means to solve the problem of the first and last mile. However, the results show that e-scooters are not primarily used for commuting but rather for leisure. This can be seen in the average distance and duration of the entire zone as well as the temporal distribution.
- The average distance and duration for the entire zone are 1.8 km and 10 min, which indicates that it is mainly very short trips that occur. Furthermore, the e-scooters are used mainly during weekday afternoons and especially during late weekend evenings when the demand for commuting is insignificant.

References

1. Röda Korset: Klimatkrisen–hög tid att hjälpa drabbade (n.d.). https://www.rodakorset.se/var-varld/klimatkrisen/. Accessed 1 Nov 2021
2. Naturvårdsverket: Sveriges klimatmål och klimatpolitiska ramverk (n.d.). https://www.naturvardsverket.se/amnesomraden/klimatomstallningen/sveriges-klimatarbete/sveriges-klimatmal-och-klimatpolitiska-ramverk/. Accessed 19 Nov 2021
3. Li, A., Gao, K., Zhao, P., Qu, X., Axhausen, K.W.: High-resolution assessment of environmental benefits of dockless bike-sharing systems based on transaction data. J. Clean. Prod. **296**, 126423 (2021)
4. Gao, K., Yang, Y., Qu, X.: Diverging effects of subjective prospect values of uncertain time and money. Commun. Transp. Res. **1**, 100007 (2021). https://doi.org/10.1016/j.commtr.2021.100007

5. Naturvårdsverket: Inrikes transporter, utsläpp av växthusgaser (n.d.). https://www.naturvardsverket.se/data-och-statistik/klimat/vaxthusgaser-utslapp-fran-inrikes-transporter/. Accessed 19 Nov 2021
6. Trafa: Resvanor i Sverige 2020 (2021). https://www.trafa.se/globalassets/statistik/resvanor/2020/resvanor-i-sverige-2020.pdf . Accessed 16 Sept 2021
7. Mansky, J.: The Motorized Scooter Boom That Hit a Century Before Dockless Scooters, Smithsonian Magazine (2019). https://www.smithsonianmag.com/history/motorized-scooter-boom-hit-century-dockless-scooters-180971989/. Accessed 3 Nov 2021
8. Region Stockholm: Elsparkcyklar i delningsekonomin – Studie med fokus på regionalutveckling (2019). http://www.rufs.se/globalassets/h.-publikationer/2020/rapport-elsparkcyklar_200305.pdf. Accessed 27 Sept 2021
9. Zhang, L., Zeng, Z., Gao, K.: A bi-level optimization framework for charging station design problem considering heterogeneous charging modes. J. Intell. Connect. Veh. **5**(1), 8–16 (2022). https://doi.org/10.1108/JICV-07-2021-0009
10. International transportation forum 2020: Safe Micromobility (2020). https://www.itf-oecd.org/sites/default/files/docs/safe-micromobility_1.pdf. Accessed 10 Oct 2021
11. Gao, K., Yang, Y., Zhang, T., Li, A., Qu, X.: Extrapolation-enhanced model for travel decision making: an ensemble machine learning approach considering behavioral theory. Knowl. Based Syst. **218**, 106882 (2021)
12. Mao, S., Xiao, G., Lee, J., Wang, L., Wang, Z., Huang, H.: Safety effects of work zone advisory systems under the intelligent connected vehicle environment: a microsimulation approach. J. Intell. Connect. Veh. **4**(1), 16–27 (2021). https://doi.org/10.1108/JICV-07-2020-0006
13. Madapur, B., Madangopal, S., Chandrashekar, M.N.: Micro-mobility infrastructure for redefining urban mobility. Eur. J. Eng. Sci. Technol. **3**(1), 71–85 (2020)
14. Shaheen, S., Cohen, A.: Shared Micromobility policy toolkit – Docked and dockless bike and scooter sharing. UC Berkeley: Transportation Sustainability Research Center, 2–7 (2019)
15. Bieliński, T., Ważna, A.: Electric scooter sharing and bike sharing user behaviour and characteristics. Sustainability **12**(22), 9640 (2020)
16. Insurance Soved Blog: Australia's electric scooter laws, Budget Direct (2021). https://www.budgetdirect.com.au/blog/australias-electric-scooter-laws-by-state.html. Accessed 2 Dec 2021
17. Transportstyrelsen: A-traktor och elsparkcykel – så många skadas (2021). https://transportstyrelsen.se/sv/Press/Pressmeddelanden/2021/a-traktor-och-elsparkcykel-sa-manga-skadas. Accessed 18 Oct 2021
18. Regionfakta: Areal och befolkningstäthet (2021). https://www.regionfakta.com/vastra-gotalands-lan/geografi/areal-och-befolkningstathet/. Accessed 20 Oct 2021
19. Lu, C., Liu, C.: Ecological control strategy for cooperative autonomous vehicle in mixed traffic considering linear stability. J. Intell. Connect. Veh. **4**(3), 115–124 (2021). https://doi.org/10.1108/JICV-08-2021-0012
20. Göteborgs stad: Statistiks databas för Göteborgs stad (n.d). http://statistikdatabasgote-borg.se/pxweb/. Accessed 20 Oct 2021
21. Göteborgs stad: Översiktsplan för Göteborgs stad (n.d.). https://oversiktsplan.goteborg.se/. Accessed 18 Oct 2021
22. Xu, Q., Li, K., Wang, J., Yuan, Q., Yang, Y., Chu, W.: The status, challenges, and trends: an interpretation of technology roadmap of intelligent and connected vehicles in China (2020). J. Intell. Connect. Veh. **5**(1), 1–7 (2022). https://doi.org/10.1108/JICV-07-2021-0010

Smart Pavement: An Attention-Based Classification Model for Road Pavement Material

Ye Yuan[1], Qingwen Xue[1], Hong Lang[1(✉)], Jie Zhu[2], Jiang Chen[1], and Peng Yuan[3]

[1] The Key Laboratory of Road and Traffic Engineering, Ministry of Education, Tongji University, Shanghai, China
honglang@tongji.edu.cn
[2] Department of Architecture and Civil Engineering, Chalmers University of Technology, 412 96 Goteborg, Sweden
[3] China Electronics Technology Group, Taiji Co. LTD., Beijing, China

Abstract. Intelligent recognition of traffic road damage is essential for realizing smart vehicles and intelligent transportation systems. The classification of road material types before recognition is a challenge for traffic road damage recognition due to differences in features such as concrete and asphalt. In addition, the widely distributed roads make environmental factors a critical factor affecting the classification. In this paper, we propose a deep learning-based road material classification method that introduces an attention mechanism to deal with the influence of different environments on road material recognition. We acquired tens of thousands of road surface images for training and testing and performed practical validation in real roads. The experiments show that our method has high accuracy and recall in road material classification.

Keywords: Machine learning · Road engineering · Intelligent transportation

1 Introduction

With the development of intelligent transportation systems, the number of vehicles operating on the road has increased dramatically, putting tremendous pressure on the existing road system. On the other hand, with the growth of smart devices, the management and the maintenance of roads also face the need for smart. Researchers have started to focus on attacking the intelligent identification of road diseases, and excellent results have been achieved [1, 2].

Road damage has different manifestations with different road surface materials. In general is divided into two main categories: asphalt and concrete. The current existence of road damage intelligent detection models is usually applied to a single road surface material [3]. However, in the existing road system, there

Y. Bie et al. (Eds.): KES-STS 2022, SIST 304, pp. 133–140, 2022.
https://doi.org/10.1007/978-981-19-2813-0_14

is a large number of different materials road mix distribution, for the deployment of intelligent equipment and intelligent transportation system has caused a significant obstacle.

The vast road network system makes manual classification of road materials costly and impractical. Moreover, due to the similar distribution of pixel values, it is not easy to distinguish different road materials automatically. Fortunately, with artificial intelligence and computer vision, a new way is provided for computers to solve this problem. In recent years, several researchers have been using artificial intelligence methods for vehicle trajectory prediction [4], vehicle self-localization [5], and etc.

There are two challenges for the classification of road materials using machine learning. First, due to the low information entropy in road images, the information contained in the images can be easily lost in operators such as convolution. Second, it is not easy to sport different materials independently because the representation of road features varies significantly in different environments or regions.

In this paper, we propose a method for road material classification based on deep learning. We apply a computer vision approach to feature extraction of road images. On the other hand, for the problem that road images vary significantly in performance in different environments, we introduce an attention module to smooth the internal information of the images.

2 Relation Work

Image classification is one of the fundamental tasks in computer vision. Researchers have persevered in it for decades, such as the maximum likelihood method, linear discriminant, etc. With the development of computer technology, neural networks, SVM, and decision trees have been developed to perform image classification.

With the development of deep learning, image classification techniques have reached new heights [6–9]. VGG networks [6] have been proposed for image classification tasks. Inception models [10] show that increasing the depth of the network can significantly improve the quality of representation learned by the network. In addition, Batch Normalization (BN) [11] regulates the distribution of inputs at each layer, adding stability to the learning process of deep networks. Wang et al. [8] proposed a novel end-to-end Attention Recurrent Convolutional Network (ARCNet) for scene classification. LSLRR [7] improves the classical low-rank representation with a location constraint criterion and a structure-preserving strategy. Building on these works, ResNets [9] demonstrated that it is possible to learn deeper and stronger networks by using identity-based skip connections. Finally following these works, some work has further reconfigured the connections between network layers [12,13], improving the learning and representation properties for deep networks.

3 Method

We propose a deep learning-based convolutional network framework applied to road material classification. The framework consists of two parts: 1. a deep residual network to preserve road features in convolutions. 2. a self-attention module to exclude the interference of the environment.

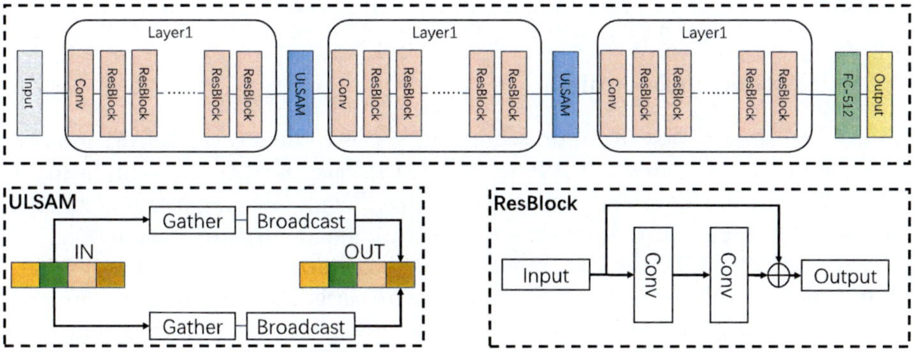

Fig. 1. The overview of our method. Our approach consists of two parts: 1. the residual network serves as a backbone network. 2. the ULSAM module is used to extract self-attention information. Given a pavement image, the residual layer encodes it into feature vectors, the feature vectors go through the ULSAM layer to calculate the internal relationships of the feature map and then send it to the next residual layer. Finally, the classification results are output through the fully connected layer.

3.1 Basic Network

We introduce residuals as the backbone network framework, called R. The basic idea of ResNets is to skip blocks of convolutional layers by using shortcut connections to form shortcut blocks (residual blocks). These stacked residual blocks greatly improve the training efficiency and the difficulty of representation learning, and largely solve the degradation problem by using BN [11] and MSR [14]. The residual network $mathcalR$ consists of a large stack of residual blocks R. The structure of each residual block is shown in the Fig. 1. Each residual block can be viewed as a mapping of Eq. 1.

$$x_{l+1} = h(x_l) + F(x_l, W_l) \tag{1}$$

where x_l and x_{l+1} are input and output of the l-th block, respectively. F is a residual mapping function, $h(x_l) = x_l$ is an identity mapping function.

The entire network is built by stacking multiple residual blocks. It has three groups of residual blocks, and each group has equal numbers of blocks. The loss function contains cross-entropy and norm Eq. 2:

$$L_{cross} = -\frac{1}{N}\sum_{n=1}^{N}\left[y_n \log\left(\overline{y_n}\right) + (1 - y_n)\log\left(1 - \overline{y_n}\right)\right], \qquad (2)$$
$$L_n = \|\overline{y} - y\|$$

where N means the class number, y_n means the class label of sample n and $\overline{y_n}$ means the probability that the sample n is predicted to be a positive class. Then $\overline{y}, y$ denote the predicted results and labels of all classes respectively.

3.2 Self-attention Module

To solve the road of feature representation, we introduce the ULSAM module [15] to build the self-attention layers. ULSAM learns individual attention maps for each feature subspace and enables compute efficient learning of cross-channel information along with multi-scale and multi-frequency feature learning.

Let $F \in R^{m \times h \times w}$ be the feature maps from an intermediate convolution layer, where m is the number of input channels, $h, andw$ is the spatial dimensions of the feature maps. ULSAM divides the input feature maps F into g mutually exclusive groups $[F_1, F_2, ...F_{\tilde{n}}, ...F_g]$ where each group have G feature maps. We define $F_{\tilde{n}}$ as a group of intermediate feature maps and proceed as Eq. 3:

$$\begin{aligned} A_{\tilde{n}} &= \text{softmax}\left(PW^1\left(\text{maxpool}^{3\times3,1}\left(DW^{1\times1}\left(F_{\tilde{n}}\right)\right)\right)\right) \\ \hat{F_{\tilde{n}}} &= \left(A_{\tilde{n}} \otimes F_{\tilde{n}}\right) \oplus F_{\tilde{n}} \\ \hat{F} &= \text{concat}\left(\left[\hat{F_1}, \hat{F_2}, \ldots.\hat{F_{\tilde{n}}}, \ldots.\hat{F_g}\right]\right) \end{aligned} \qquad (3)$$

In the Eq, $\text{maxpool}^{3\times3,1}$ is maxpool with kernel size $= 3 \times 3$ and padding $= 1$, $DW^{1\times1}$ is depthwise convolution with 1×1 kernel, PW^1 is pointwise convolution with only one filter, and $A_{\tilde{n}}$ is an attention map inferred from a group of intermediate feature maps $F_{\tilde{n}}$. Attention map $A_{\tilde{n}}$ in each group (subspace) captures the non-linear dependencies among the feature maps by learning to gather cross channel information. ULSAM employ a gating mechanism with a softmax activation. Each group of feature maps gets the refined set of feature maps $F_{\tilde{n}}$ after the featurere distribution where $\oplus$ is element-wise multiplication and $\oplus$ is element-wise addition. The final output of ULSAM $\hat{F}$ is obtained by concatenating the feature maps from each group.

The multi-scale feature learning of ULSAM can provide accurate representation of pavement image features and avoid feature interference in different environments.

4 Experiments

4.1 Datasets

We collected images of all classes of roads throughout China as a dataset. A total of 23122 images containing 2892 concrete roads and 20230 asphalt roads were included, the samples as shown in Fig. 2. We capture 1 mm pavement images using a vehicle equipped with a line laser scanner. We use 3D line laser scanners to obtain all data with a sampling accuracy of 1 mm mounted on the engineering vehicle. The raw data is obtained by scanning the road information as the vehicle moves. We then converted the raw data to image format and solved the varying brightness of the laser-acquired images by homogenizing the light operation. The original size of the image was 2048×2048 and was resized to 512×512 for input. We take 80% as the test set, 10% as the validation set, and 10% as the test set. The accuracy reported on the Road datasets is obtained by training the models from scratch (not on the ImageNet pre-trained model).

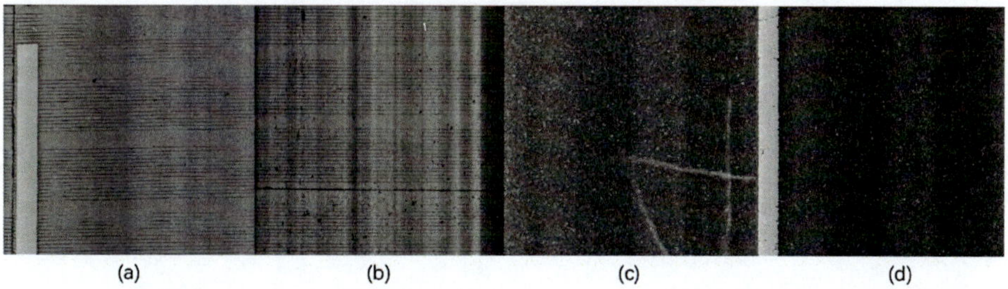

Fig. 2. The samples of dataset. (a) and (b) show images of concrete pavement. (c) and (d) show images of asphalt pavement.

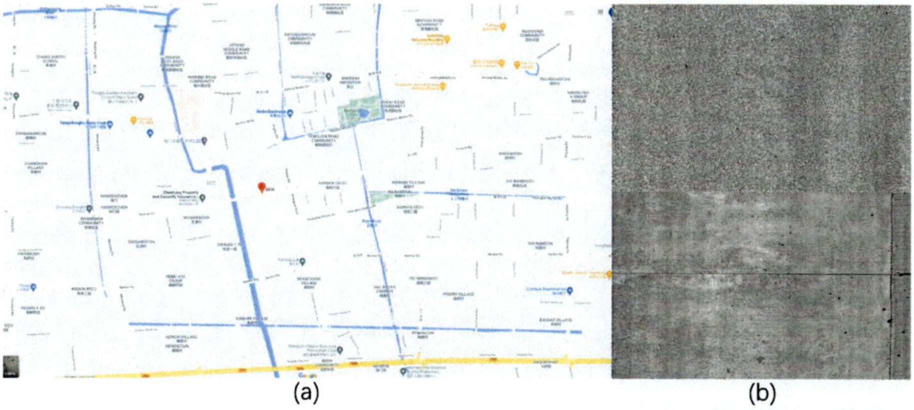

Fig. 3. The location of the case study and sample images.

4.2 Experimental Setting

We perform experiments using PyTorch deep learning framework. We used four 1080Ti GPU for experiment. We train all method with batch size 4 and adam optimizer for 200 epochs. The initial learning rate is set to 0.001.

4.3 Comparison to Existing Methods

In order to accurately benchmark the performance of our method, we compared it with: (1) ResNet20 (20 means number of resblock) (2) ResNet32 (3) ResNet44 (5) ResNet56 (6) Our(bone network based on ResNet20). All comparison methods were set up the same as in the 4.2 section to ensure unbiased results of the experiment.

Table 1. Comparison results with other methods

Methods	Precision
ResNet20	96.078
ResNet32	98.039
ResNet44	99.010
ResNet56	99.505
Our	99.846

As shown in Table 1, our method achieves the highest precision rate performance. The improved precision rate of 3.768 compared to Bone network (resnet20) is even better than other computational space-consuming methods.

4.4 Case Study

To further measure the effectiveness of our method, we searched for two road with different environments in different regions to test our method.

Setting. We chose a section of a high-grade road with mixed pavement materials in Haimen, Jiangsu, China, at around 10:00 a.m. BST on April 14, 2021, as a case study. There are 1727 pavement images of the road, including 804 images of concrete pavement and 922 images of asphalt pavement, as shown Fig. 3. We validate our method performance using the model trained in the dataset and the data collected in the case. All training settings are the same as in the 4.2 section.

Results. As shown in Table 2, our model exhibits very high robust performance, and the practical implementation maintains stability in terms of precision and recall relative to the laboratory setting.

Table 2. Case study metrics

Classes	Results	Ground Truth
Concrete	802	804
ResNet32	925	922
Classes	Results	Precision
Wrong concrete	1	99.88
Wrong asphalt	3	99.68
Classes	Results	Recall
Wrong concrete	1	99.88
Wrong asphalt	3	99.67

5 Discussion

We propose a deep learning-based road material classifier method that introduces an attention mechanism to optimize road feature extraction for different environments. The performance of our model achieves a significant accuracy rate. Our future work will exploit different applications of the attention mechanism to capture complex global relationships in complex road systems to help the development of smart vehicles and intelligent transportation systems.

Acknowledgement. This work was supported by the National Natural Science Foundation of China: [Grant Number NSFC71871165] and the Fundamental Research Funds for the Central Universities: [Grant Number TTS2021-03], TaiJi chason machine learning engine.

References

1. Lang, H., Lu, J.J., Lou, Y., Chen, S.: Pavement cracking detection and classification based on 3D image using multiscale clustering model. J. Comput. Civ. Eng. **34**(5), 04020034 (2020)
2. Wen, T., et al.: Automated pavement distress segmentation on asphalt surfaces using a deep learning network. Int. J. Pavement Eng., 1–14 (2022)
3. Lang, H., Lou, Y., Lu, J.J., Chen, Y.: A review of 3D pavement automatic measurement system. In: CICTP 2017: Transportation Reform and Change-Equity, Inclusiveness, Sharing, and Innovation, pp. 831–841 (2018)
4. Shi, X., Zhao, D., Yao, H., Li, X., Hale, D.K., Ghiasi, A.: Video-based trajectory extraction with deep learning for high-granularity highway simulation (HIGH-SIM). Commun. Transp. Res. **1**, 100014 (2021)
5. Jiang, Z., Xu, Z., Li, Y., Min, H., Zhou, J.: Precise vehicle ego-localization using feature matching of pavement images. J. Intell. Connected Veh. (2020)
6. Simonyan, K., Zisserman, A.: Very deep convolutional networks for large-scale image recognition. arXiv preprint arXiv:1409.1556 (2014)

7. Wang, Q., He, X., Li, X.: Locality and structure regularized low rank representation for hyperspectral image classification. IEEE Trans. Geosci. Remote Sens. **57**(2), 911–923 (2018)
8. Wang, Q., Liu, S., Chanussot, J., Li, X.: Scene classification with recurrent attention of VHR remote sensing images. IEEE Trans. Geosci. Remote Sens. **57**(2), 1155–1167, 100014 (2018)
9. He, K., Zhang, X., Ren, S., Sun, J.: Deep residual learning for image recognition. In: Proceedings of the IEEE Conference on Computer Vision and Pattern Recognition, pp. 770–778 (2016)
10. Szegedy, C., et al.: Going deeper with convolutions. In: Proceedings of the IEEE Conference on Computer Vision and Pattern Recognition, pp. 1–9 (2015)
11. Ioffe, S., Szegedy, C.: Batch normalization: accelerating deep network training by reducing internal covariate shift. In: International Conference on Machine Learning, pp. 448–456 (2015). PMLR
12. Huang, G., Liu, Z., Van Der Maaten, L., Weinberger, K.Q.: Densely connected convolutional networks. In: Proceedings of the IEEE Conference on Computer Vision and Pattern Recognition, pp. 4700–4708 (2017)
13. Chen, Y., Li, J., Xiao, H., Jin, X., Yan, S., Feng, J.: Dual path networks. arXiv preprint arXiv:1707.01629 (2017)
14. He, K., Zhang, X., Ren, S., Sun, J.: Delving deep into rectifiers: surpassing human-level performance on ImageNet classification. In: Proceedings of the IEEE International Conference on Computer Vision, pp. 1026–1034 (2015)
15. Saini, R., Jha, N.K., Das, B., Mittal, S., Mohan, C.K.: ULSAM: ultra-lightweight subspace attention module for compact convolutional neural networks. In: Proceedings of the IEEE/CVF Winter Conference on Applications of Computer Vision, pp. 1627–1636 (2020)

Traffic Flow Model of the Weaving Section in Signalized Roundabouts

Tianshu Zhan[1], Xianmin Song[1(✉)], Yunxiang Zhang[1], and Kunwei Wang[2]

[1] School of Transportation, Jilin University, Changchun 130022, Jilin, China
songxm@jlu.edu.cn

[2] Qingdao Hisense Network Technology Co., Qingdao 266071, Shandong, China

Abstract. The efficiency of the roundabout is severely constrained by the presence of multiple conflicting traffic flows in the weaving section. An accurate description of the traffic flow in the weaving section of roundabouts is the key to intelligent traffic control. In this paper, the approach, the circulation section, the weaving section and the exit of roundabouts are taken as a unit to analyze, and a model of traffic flow for entering the weaving section is established. The impact of entering flow, circulation flow and leaving flow on the condition of the weaving section is analyzed and a model of traffic flow for leaving the weaving section is established. A new traffic flow model of the weaving section is also established considering an iterative-algorithm-based parameter calibration method. Finally, this paper builds a VISSIM-based simulation platform and collects field data to verify the accuracy of the traffic flow model. The results show that the average relative error of the simulation outputs and the values of theoretical model calculation is within 10%, and with the increase of passing vehicles, the calculation error decreases. Therefore, the model established in this paper can accurately calculate the traffic flow characteristics of the weaving section in roundabouts, which can provide theoretical support for the microscopic simulation of roundabouts and the formulation of traffic management strategies.

Keywords: Roundabout · Traffic flow model · Weaving section · Entering flow · Leaving flow

1 Introduction

Traffic flow theory mainly describes the operating characteristics of vehicles and pedestrians and expounds the generation mechanism of traffic phenomenon so as to find the cause of traffic congestion, improve the utilization efficiency of road facilities, and provide theoretical support for urban traffic management [1–3]. Probability method was widely used in early studies on traffic flow theory. Kinzert et al. first proposed the possibility of studying traffic flow with Poisson distribution [4]. Greenshields et al. used Poisson distribution to analyze the traffic condition in intersections and established the speed-density linear expression [5]. By the 1950s, the probability method was gradually replaced by many new findings, mainly including the traffic wave theory and the

Y. Bie et al. (Eds.): KES-STS 2022, SIST 304, pp. 141–150, 2022.
https://doi.org/10.1007/978-981-19-2813-0_15

car following theory [6]. Based on machine learning, Yang Liu et al. proposed deep learning architectures, termed as Deep Traffic State Prediction (DeepTSP) [7] and Deep Passenger Flow (DeepPF) [8], to predict traffic flows. Shuaian Wang et al. proposed a holistic modeling framework based on the concept of mean absolute error minimization to empirically calibrate stochastic traffic flow fundamental diagram [9].

The traffic flow theory has been carefully studied by scholars. As the key nodes of urban road networks, roundabouts have also been studied. Fouladvand et al. analyzed the delay of single-lane roundabouts and established a key traffic flow value model [10]. Lakouari et al. proposed a stochastic cellular automata model to study the traffic behavior in a single-lane roundabout and got five distinct traffic phases, namely free flow, congestion, maximum current, jammed, and gridlock [11]. A metering signal-based strategy was examined by Yuzhou Duan et al. to mitigate the problems of large entry delays that are often observed in roundabouts with unbalanced flow patterns [12]. Davidovic et al. analyzed the speed characteristics of traffic flow in urban roundabouts with different geometrical characteristics in the city of Banja Luka and established the correlation between the achieved speeds and geometrical characteristics of the intersection [13]. By analyzing the traffic capacities and traffic behaviors of domestic and foreign weaving areas and combining them with field investigation, Haijian Li et al. proposed the active and fine lane management methods for ICVs to optimal driving behavior in a weaving area [14]. Huang divided the weaving section of the single-lane roundabout into one-dimensional rectangular cells, and an asymmetric simple exclusion process was used to simulate the vehicle operation process. Therefore, Huang adopted the phase changes of entering and leaving roundabouts [15] and the phase transitions rules in single-line roundabouts [16].

Most of the existing studies on the traffic characteristics of roundabouts mainly focus on single-lane roundabouts. However, there are few studies about operational characteristics of traffic flow in multi-lane roundabouts. Considering that the capacity of an approach in a multi-lane roundabout is closely related to the traffic flow of relevant circulation lanes and the flow of relevant weaving sections, and the average speed of the vehicles in the weaving section is related to the number of vehicles and weaving behavior, it is necessary to establish a model of traffic flow for multi-lane roundabouts, which can provide theoretical support for the management of roundabouts.

2 Model Development

2.1 Description

The paper selects *multi-lane* roundabouts as the study object, as is shown in Fig. 1. Due to the special geometric characteristics of the roundabouts, vehicles have the same operating rules in each circulation section (CS) and weaving section (WS). Therefore, the CS hr, the WS r and the corresponding approach i are selected as study objects.

During the establishment process of the model in this paper, the impact of the vehicles in the CS and the WS on vehicle operation of the approach is considered. Therefore, the following assumptions must be met during the modeling:

(i) Follow the rule of "vehicles in roundabouts first", i.e. entering vehicles give way to vehicles in roundabouts.
(ii) An exclusive right-turn lane is provided within the roundabout, i.e. right-turning vehicles do not enter the WS, and therefore vehicles in the WS will not be impacted.

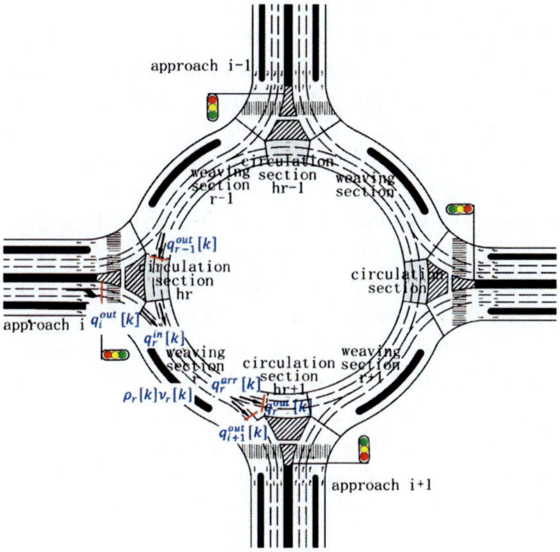

Fig. 1. Structure of the roundabout.

2.2 Model Methodology

Efforts are made in this section to analyze the traffic characteristics of the roundabout and establish a model to describe the relationship among the maximum release rate of the approach and the arrival rate of the CS, and the capacity of the WS. We also propose a model to describe the relationship between the average speed of the vehicles and the flow of the WS, and the ratio of entering to leaving. Based on the spatial-temporal transition mechanism of the vehicles in roundabouts, a traffic flow model of the WS is established, which can reflect the vehicle operating condition in roundabouts.

The traffic flow model consists of three parts: the model of traffic flow for entering the WS, the model of traffic flow for leaving the WS and the parameter calibration method.

The Model of Traffic Flow for Entering the WS. The entering flow of the WS consists of the entering flow from the CS and the approach flow. Therefore, it can be calculated as follows.

$$q_r^{in}[k] = q_{r-1}^{out}[k] + q_i^{out}[k] \tag{1}$$

where $q_r^{in}[k]$ is the number of vehicles entering the upstream of the WS r during step k, $pcu/\Delta t\ s$; $q_{r-1}^{out}[k]$ is the number of vehicles leaving the WS $r-1$ during step k,

$pcu/\Delta t\ s$; $q_i^{out}[k]$ is the number of vehicles leaving the approach i during step k, $pcu/\Delta t\ s$; k is the time step, an index corresponding to time, $t = k\,\Delta\,t$.

As the principle of "vehicles in roundabouts first", when there are few vehicles in the approach i, vehicles can enter the WS and the queuing length in the approach will not exceed a certain value. Otherwise, the queuing length in the approach will gradually increase and the queue-jump behavior of vehicles in the approach will severely impact the operation of vehicles in the roundabout and may lead to a gridlock phenomenon in the roundabout. The number of vehicles released from the approach can be calculated as follows, providing that the normal operation of traffic in the roundabout is guaranteed.

$$q_i^{out}[k] = min\{q_i^{arr}[k] + x_i[k], S_i \cdot g_i[k]\} \tag{2}$$

where $q_i^{arr}[k]$ is the number of vehicles arriving at the end of the queue in the approach i during step $k, pcu/\Delta ts$; $x_i[k]$ is the number of queuing vehicles in the approach i during step $k, pcu/\Delta t\ s$; S_i is the maximum release rate of the approach i during step $k, pcu/h$; $g_i[k]$ is the green time of the approach i during step k, h.

In order to establish a model to describe the relationship between the maximum release rate of the approach, the number of vehicles in the WS and the number of vehicles entering the CS, a VISSIM-based simulation platform is built.

The number of vehicles passing the WS is collected by varying the arriving flow in the approach, providing that the number of vehicles arriving the CS is guaranteed. The relationship between q_i^{arr}, q_{r-1}^{out} and q_r^{in} is shown in Fig. 2 as follows:

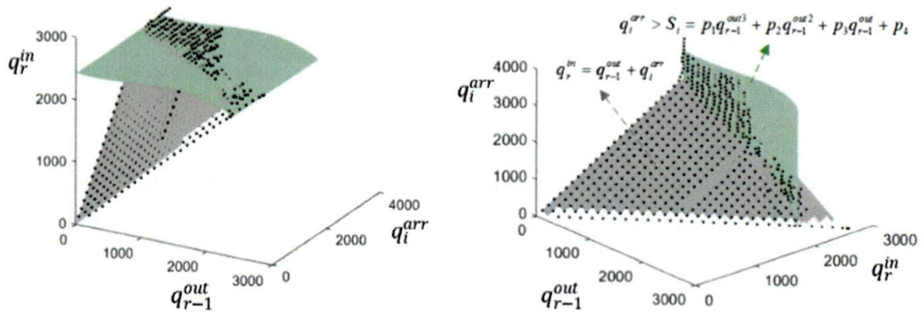

Fig. 2. Relationship among q_i^{arr}, q_{r-1}^{out} and q_r^{in}.

As is shown in Fig. 2, when the sum number of arriving vehicles from the approach and the CS is less than the capacity of the WS, the number of passing vehicles in the WS is the sum number of the arriving vehicles from the approach and the CS. When it is above the capacity of the WS, the number of arriving vehicles will gradually increase until the number of passing vehicles in the WS is roughly static, providing that the number of arriving vehicles is a certain value. Therefore, we can draw the conclusion that the number of vehicles entering the WS from the approach has no relationship with the number of vehicles in the approach. The maximum release rate of the approach can be calculated as follows:

$$S_i = p_1 {q_{r-1}^{out}}^3 + p_2 {q_{r-1}^{out}}^2 + p_3 q_{r-1}^{out} + p_4 \quad (3)$$

where p_1, p_2, p_3, p_4 are model parameters to be calibrated.

Therefore, the number of vehicles entering the WS can be calculated as follows:

$$q_r^{in}[k] = q_{r-1}^{out}[k] + min\left\{ q_i^{arr}[k] + x_i[k], \left(p_1 {q_{r-1}^{out}}^3 + p_2 {q_{r-1}^{out}}^2 + p_3 q_{r-1}^{out} + p_4 \right) g_i[k] \right\} \quad (4)$$

The Model of Traffic Flow for Leaving the WS. By analyzing the travelling characteristics of vehicles in roundabouts, we can draw the conclusion that only when the observing step exceeds the time required for vehicles to pass the WS at free-flow speed can leaving vehicles be observed at the end of the WS.

Therefore, the number of vehicles leaving the WS can be calculated as follows:

$$q_r^{arr}[k] = \begin{cases} min\{\rho_r[k] \cdot v_r[k] \cdot n_r \cdot \Delta t, N_r\} & if \;\; k \geq \frac{l_r}{\Delta t \cdot v_r^{free}} \\ 0 & otherwise \end{cases} \quad (5)$$

where $q_r^{arr}[k]$ is the number of vehicles arriving the downstream of the WS $r, pcu/\Delta t\, s$; $\rho_r[k]$ is the density of vehicle flow in the WS r during step $k, pcu/km/lane$; $v_r[k]$ is the average speed of traffic flow in the WS r during step k, km/h; n_r is the number of lanes in the WS r; Δt is system update interval, s; N_r is the capacity of the WS r, pcu; l_r is the average length of lanes in the WS r, km; v_r^{free} is the free-flow speed in the WS r, km/h.

Entering vehicles need to find gaps to enter the WS and leaving vehicles also need to change lanes in time to leave the WS. Therefore, vehicles in the WS need to slow down to avoid the above two types of vehicles, and the average speed of traffic flow in the WS will be impacted. We use the SPSS to analyze the correlation between the average speed of traffic flow in the WS and the ratio of entering to leaving. The result shows that $R^2 = 0.8037$, which means the ratio of entering to leaving has a significant impact on the average speed of traffic flow in the WS. Also, the average speed of traffic flow in the WS is impacted by the number of passing vehicles.

The relationship among the average speed of vehicles in the WS, the ratio of entering to leaving, and the number of passing vehicles are shown in Fig. 3.

As is shown in Fig. 3, vehicles in the WS travel at speed close to the free-flow speed when the number of passing vehicles is below 500 pcu/h, and as the number of passing vehicles and ratio of entering to leaving increase, the average speed of vehicles in the WS gradually decreases until it is close to the jam-speed. Therefore, the average speed of vehicles in the WS can be calculated as follows:

$$v_r[k] = \begin{cases} v_r^{free} & q_r[k] < q_r^{min} \&\& \alpha_r[k] < \alpha_r^{min} \\ v_r^{min} & q_r[k] > q_r^{max} \&\& \alpha_r[k] > \alpha_r^{max} \\ \omega_1 + \omega_2 q_r[k] + \omega_3 \alpha_r[k] + \omega_4 q_r[k]^2 + \omega_5 q_r[k] \cdot \alpha_r[k] & otherwise \end{cases} \quad (6)$$

where v_r^{min} is the jam-flow speed in the WS $r, km/h$; $\omega_1, \omega_2, \omega_3, \omega_4$ are model parameters to be calibrated; $q_r[k]$ is the number of vehicles passing the WS r during step $k, pcu/\Delta t\; s$; $q_r^{min}\left(q_r^{max}\right)$ is the flow of vehicles when traveling at free-flow speed (jam-flow speed) in the WS $r, pcu/h$; $\alpha_r[k]$ is the ratio of the number of vehicles entering and leaving the

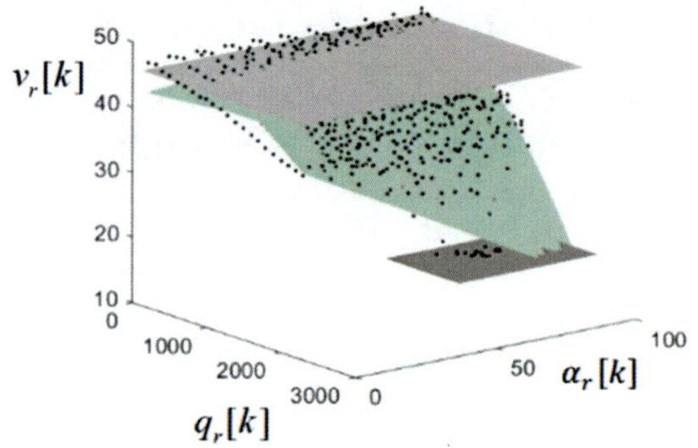

Fig. 3. Relationship among $q_r[k]$, $v_r[k]$, *and* $\alpha_r[k]$

WS r to the total number of the vehicle during step k; $\alpha_r^{min}(\alpha_r^{max})$ is ratio of entering to leaving when traveling at free-flow speed (jam-flow speed) in the WS r.

The number of passing vehicles in the WS r is the sum of the number of entering vehicles during the step k and the number of present vehicles at the initial moment of step k. It can be calculated as:

$$q_r[k] = q_r^{in}[k] + N_r[k] \tag{7}$$

where $N_r[k]$ is the number of vehicles in the WS r at the initial moment of step k.

The ratio of entering to leaving in the WS r can be calculated as:

$$\alpha_r[k] = \varepsilon_r[k] + \chi_r[k] \tag{8}$$

where $\varepsilon_r[k]$ is the ratio of number of vehicles entering the WS r from the CS hr and leaving from the exit lane $i+1$ to the total number of vehicles passing the WS r during step k, assuming that we know $\varepsilon_r[0]$ at the initial moment; $\chi_r[k]$ is the ratio of the number of vehicles entering the WS r from the approach i to the total number of vehicles passing the WS r during step k, assuming that we know $\chi_r[0]$ at the initial moment.

The number of vehicles in the WS can be calculated by substituting the Eq. (7) and (8) into the Eq. (6) and substituting the Eq. (6) into the Eq. (5). According to the different destinations, the vehicles leaving the WS r can be de divided into the vehicles entering the CS $r+1$ and the vehicles arriving the exit $i+1$, which can be described by Eq. (9).

$$q_r^{arr}[k] = q_r^{out}[k] + q_{i+1}^{out}[k] \tag{9}$$

where:

$$q_r^{out}[k] = (\gamma_r[k] + \chi_r[k]) \cdot q_r^{arr}[k] \tag{10}$$

$$q_{i+1}^{out}[k] = \varepsilon_r[k] \cdot q_r^{arr}[k] \tag{11}$$

Parameter Calibration. The entering and leaving of vehicles in the WS will lead to the change of the total number of vehicles and the proportion of flow distribution in the WS. Therefore, an iterative-algorithm-based parameter calibration method should be established to make sure the continuity of the iterative process of the traffic model.

The iterative process of the total number of vehicles in the WS r is as follows:

$$N_r[k+1] = N_r[k] + \sum\nolimits_{j \in \Gamma(r)} Q_{jr}[k] - \sum\nolimits_{j \in \Gamma^{-1}(r)} Q_{rj}[k] \quad (12)$$

where $\sum_{j \in \Gamma(r)} Q_{jr}[k]$ is the total number of vehicles entering the WS r from upstream approach (or the CS) j during step $k, pcu/h$; $\sum_{j \in \Gamma^{-1}(r)} Q_{rj}[k]$ is the total number of vehicles travelling from the WS r to downstream exit (or circulation section) j during step $k, pcu/h$.

The iterative process of the distribution rate in the WS r is as follows:

$$\gamma_r[k+1] = \left[\gamma_r[k] \cdot N_r[k+1] + q_i^{out}[k+1] \cdot \beta_r[k+1]\right] / q_r[k+1] \quad (13)$$

$$\varepsilon_r[k+1] = \left[\varepsilon_r[k] \cdot N_r[k+1] + q_{r-1}^{out}[k+1] \cdot (1 - \beta_r[k+1])\right] / q_r[k+1] \quad (14)$$

$$\chi_r[k+1] = \left(\chi_r[k] \cdot N_r[k+1] + q_i^{out}[k+1]\right) / q_r[k+1] \quad (15)$$

where $\gamma_r[k]$ is the ratio of the number of vehicles entering the CS hr from the WS r and passing the WS r and leaving the CS $hr + 1$ to the number of the total vehicles passing the WS r during step k; $\beta_r[k]$ is the ratio of the number of vehicles entering the CS hr to leaving the circulation area $hr + 1$ and the number of vehicles entering the WS r during step k.

3 Model Validation

In order to verify the accuracy of the traffic flow models above, we use VISSIM to analyze the field traffic flow characteristics of a case roundabout, and evaluate the accuracy of the models by comparing the simulation outputs and the values calculated by the theoretical models.

3.1 Simulation Condition Setting

The WS between the Zhengyang Street and the Jinyang Avenue in Jinyang Square can meet the simulation condition and is chosen as our study object, as is shown in Fig. 4 (left).

In Fig. 4 (left), $\alpha = 74.70°$, $L = 84.8$ m. In the WS, the travel time of 1351 passing vehicles in 2 h is collected and the average speed of vehicles is calculated. We plot a cumulative velocity curve, as is shown in Fig. 4 (right). The obtained speed distribution patterns are used in the speed distribution of VISSIM.

The parameters of the geometric condition, vehicle type, vehicle behavior, driving behavior, and the survey interval are calibrated in Table 1.

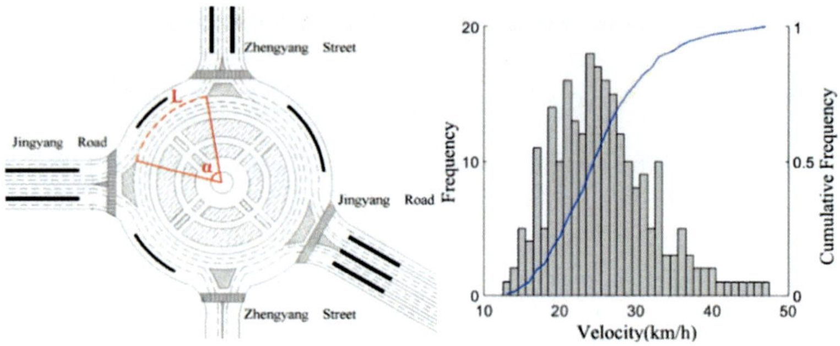

Fig. 4. Structure of Jingyang Square (left) and Velocity distribution in the WS (right)

Table 1. Other simulation condition settings.

Index	Settings	Font size and style
Geometric design	The length of WS (L) (m) The width of WS (W) (m) The width of approach $(e1)$ (m) The width of CS $(e2)$ (m) The average width of WS (e) (m)	84.8 14 8 10.5 10.5
Driving behavior	Using the car following model of Wiedemann74 Additional safety distance (m) Increased safety distance available (m)	2.75 3.75
Interval	Simulation cycle Sampling interval	1 h (3600 s) 5 min

3.2 Parameter Calibration

The detectors are set at the corresponding sections as the red line shown in Fig. 1. 2163 types of flow operation characteristics are obtained by varying the lane changing ratio and arriving flow in the approach and CS. The result of parameter calibration is shown in Table 2.

Table 2. The results of parameter calibration.

Parm	Calibration result	R^2
S_i	$S_i = -0.0009q_{r-1}^{out2} + 0.2166q_{r-1}^{out} + 2280.4$	0.7953
$v_r[k]$	$v_r(k) = 40.8\, while\, q_r[k] < 400 \,\&\&\, \alpha_r[k] < 80\%$ $v_r[k] = 27.28 - 0.009896q_r[k] - 0.1251\alpha_r[k] + 0.000004684q_r[k]^2 + 0.0002623q_r[k] \cdot \alpha_r[k]$ $while\, q_r[k] \in [400, 1500) \| \alpha_r[k] \in [70\%, 80\%)$ $v_r[k] = 13.5\, while\, q_r[k] > 1500 \&\& \alpha_r[k] > 80\%$	0.7620 0.8823 0.88826

3.3 Result Analysis

The number of vehicles leaving the WS can be calculated by substituting the results in Table 2 into the traffic flow model. By comparing the calculation results and the simulation outputs, the figures of the calculation error and the error percentage can be plotted, as is shown in Fig. 5.

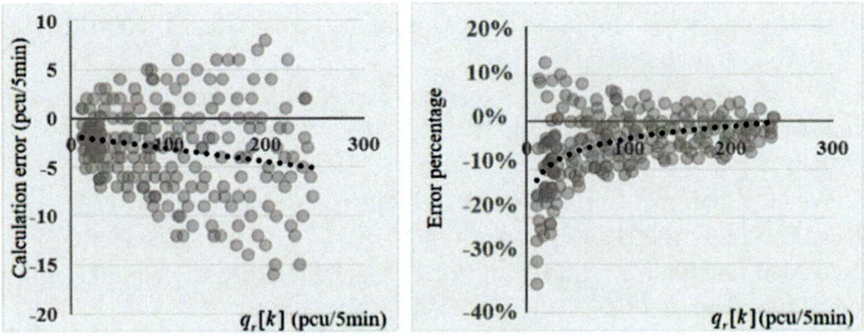

Fig. 5. The results of simulation

As is shown in Fig. 5, when the number of passing vehicles is 208 pcu/5min, the maximum calculation error of the model is 16 pcu/5min, and the average relative error percentage is 7.96%, which is within the range of acceptable error, which means that the traffic flow characteristics of the WS in roundabouts can be accurately described.

4 Conclusions

The paper analyzes the relationship among the number of vehicles passing the WS and the flow of the approach and the CS and establishes a traffic flow model of the WS in signalized roundabouts. The relationships among the maximum release rate of the approach, the arriving flow, the travel speed of CS and other factors are described by the model. In the simulation section, the average relative error is controlled within 10%, which means the model established in the paper can provide theoretical support for the management of signalized roundabouts and the implementation of simulation software.

Acknowledgments. This study is supported by the program *A Signalized Control Method for Preventing Intersection Gridlock Based-Video Processing* (No. 20190201107JC) of the Natural Science Foundation of Jilin Province.

References

1. Jiang, R., Wu, Q., Zhu, Z.: Full velocity difference model for a car-following theory. Phys. Rev. E **64**, 017101 (2001)
2. Biham, O., Middleton, A., Levine, D.: Self-organization and a dynamical transition in traffic-flow models. Phys. Rev. A **46**(10), 6124–6127 (1992)

3. Jiang, R., Wu, Q., Zhu, Z.: A new continuum model for traffic flow and numerical tests. Transp. Res. Part B Methodol. **36**(5), 405–419 (2002)
4. Kinzer, J.P.: Application of the theory of probability to problems of highway traffic. Dissertation, Polytechnic Institute of Brooklyn (1993)
5. Greenshields, A.: Study of traffic capacity. Proc. Highw. Res. Board **14**, 448–477 (1934)
6. Robert, E.C., Robert, H., Elliott, W.M.: Traffic dynamics: studies in car following. Oper. Res. **6**(2), 165–184 (1958)
7. Liu, Y., Lyu, C., Zhang, Y., Liu, Z., Yu, W., Qu, X.: DeepTSP: deep traffic state prediction model based on large-scale empirical data. Commun. Transp. Res. **1**, 100012 (2021). https://doi.org/10.1016/j.commtr.2021.100012
8. Liu, Y., Liu, Z., Jia, R.: DeepPF: a deep learning based architecture for metro passenger flow prediction. Transp. Res. Part C Emerg. Technol. **101**, 18–34 (2019). https://doi.org/10.1016/j.trc.2019.01.027
9. Wang, Y., Wu, J., Chen, K., Liu, P.: Are shared electric scooters energy efficient? Commun. Transp. Res. **1**, 100022 (2021). https://doi.org/10.1016/j.commtr.2021.100022
10. Fouladvand, M.E., Sadjadi, Z., Shaebani, M.R.: Characteristics of vehicular traffic flow at a roundabout. Phys. Rev. E **70**(2), 046132 (2004)
11. Lakouari, N., Oubram, O., Ez-Zahraouy, H., Cisneros-Villalobos, L.: Traffic flow behavior at a single-lane urban roundabout. Int. J. Mod. Phys. C **29**(3), 1 (2018)
12. Duan, Y., Qu, X., Easa, S., Yan, Y.: Optimising total entry delay at roundabouts with unbalanced flow: a dynamic strategy for smart metering. IET Intel. Transport Syst. **13**(3), 485–494 (2019)
13. Davidovic, S., Bogdanovic, V., Garunovic, N., Papic, Z., Pamucar, D.: Research on speeds in roundabouts for the needs of sustainable traffic management. Sustainability **13**(1), 399 (2021)
14. Li, H., Zhang, J., Zhang, Z., Huang, Z.: Active lane management for intelligent connected vehicles in weaving areas of urban expressway. J. Intell. Connect. Veh. **4**(2), 52–67 (2021). https://doi.org/10.1108/JICV-08-2020-0009
15. Huang, D.W.: Phase diagram of a traffic roundabout. Physica A **383**(2), 603–612 (2007)
16. Huang, D.W.: Modeling gridlock at roundabout. Comput. Phys. Commun. **189**, 72–76 (2015)

On the Impact Analysis of Emergency Vehicles Preemption on Signalized Intersections with Connected Vehicles

Jian Xie[1], Jiaming Wu[2](✉), and Runkai Yang[1]

[1] Department of Traffic Management, People's Public Security University of China, Beijing 100038, China
[2] Department of Architecture and Civil Engineering, Chalmers University of Technology, Gothenburg, Sweden
Jiaming.wu@chalmers.se

Abstract. In recent years, vehicle-to-infrastructure (V2I) and vehicle-to-vehicle (V2V) communication are the leading technologies in intelligent transportation systems. With V2I and V2V, new strategies are enabled to better facilitate emergency rescue and traffic emergency management. Since signalized intersection is the major source of delay and congestion, how to prioritize emergency vehicles (EVs) at intersections becomes an indispensable research topic and has drawn extensive studies. Albeit that emergency vehicles should be granted absolute priority, the influence of EVs on the intersection was rarely studied, especially when aggressive preemption strategies are adopted. To this end, this paper is devoted to evaluating the impacts of EVs on the control of signalized intersections, as an effort to better emergency management. Mixed-integer linear programming models are developed to optimize the intersection control before and after the pass of EVs, respectively. Case studies indicate that, by assuming the EV will take exclusively one lane at the intersection, the average delay could increase by 11.28% along with a capacity decrease of 6.63% .

Keywords: Emergency vehicle · Connected vehicles · Intersection control

1 Introduction

With the rapid development of urbanization, urban transportation systems are becoming increasingly congested, bringing a series of side effects such as air pollutions, noise, and traffic accidents. Traffic accidents, in particular, endanger the life and property of the general public. Each incident poses a challenge to the emergency management capacity of the transport agency. According to research, half of the serious injuries can be treated if treated in a timely manner within 30 min of an accident. In medical science, the 30 min are called "prime time" [1].

In traditional intersections, emergency vehicles cannot have access to real-time information, such as queue length and signal, which hinders the key decision-making of speed control and route choices. It is common that a large queue exists at the intersection when

Y. Bie et al. (Eds.): KES-STS 2022, SIST 304, pp. 151–160, 2022.
https://doi.org/10.1007/978-981-19-2813-0_16

Fig. 1. Emergency vehicle travel in smart connected environment

emergency vehicles arrive due to the lack of communication and cooperation. In the environment of the smart connected vehicles, real-time sharing of vehicles and road information, can be realized through V2V and V2I to enable smarter EV preemption as well as intersection control.

In this area, a large body of research focused on the development of emergency vehicle priority systems. An emergency vehicle priority system ensures the passage of emergency vehicles through priority signal control, route guidance and social vehicle diversion [2]. The use of TPNS in the modeling emergency vehicle priority system was studied to ensure that emergency vehicles can pass through intersections quickly [3]. There are also dynamic preemption strategies that tend to balance different objectives [4].

In recent years, Intelligent Connected Technology is also of great significance to emergency rescue and traffic emergency management control. Intersection signal coordination optimization, enabled by smart connected technology, can accurately obtain the vehicle trajectory information and feed it to the controller for better optimization performance. Intersection collisions are studied in the context of CVIC, so that vehicles can safely and quickly pass through the intersection without collision with other vehicles [5–7]. A lane advance strategy based on integer linear programming (ILP) algorithm was proposed to achieve the absolute priority of emergency vehicles by clearing lanes for approaching emergency vehicles [8].

In order to solve the problem of low traffic efficiency at intersections, two experimental scenarios were designed, using intelligent networked vehicles to guide multiple ordinary vehicles [9]. A * algorithm is used to solve the intelligent interconnected vehicle sequencing problem and find the shortest path quickly [10]. The influence of whether to force lane change on execution time in intelligent networked vehicle environment was discussed [11].

Most of the current research on emergency vehicle preemption gives priority to emergency vehicles based on a limited number of road detectors. The detectors are however costly, cumbersome to set up, and unreliable, especially under unfavorable weather. At the same time, under the control environment of traditional intersection,emergency vehicles cannot have access to the exact information of intersection, such as queue length, traffic information. Frequent acceleration and deceleration at crowded urban intersections make it even more challenging to dissipate queues when emergency vehicles arrive.

This to some extent leads to aggressive preemption strategies, even with connected and automated vehicles [12, 13]. However, few studies investigated the influence of such preemption strategies. To this end, the present paper is devoted to evaluating the influence of EV preemption strategies on the intersection performance under optimal control.

2 Methodology

In this study, we assume to grant the emergency vehicles the highest priority so that they can pass through the intersection without delay. In the smart connect environment, a lane at an intersection can be cleared and used exclusively for EVs. When a lane is occupied by emergency vehicles, as shown in Fig. 1, the signal control plan is adjusted to optimize the intersection capacity for normal vehicles. In summary, we seek to evaluate the impacts on signalized intersections while ensuring the priority of emergency vehicles. The signal timing optimization and the collaborative control of the emergency vehicle are coordinated and modeled in a mixed-integer linear programming framework (Fig. 2).

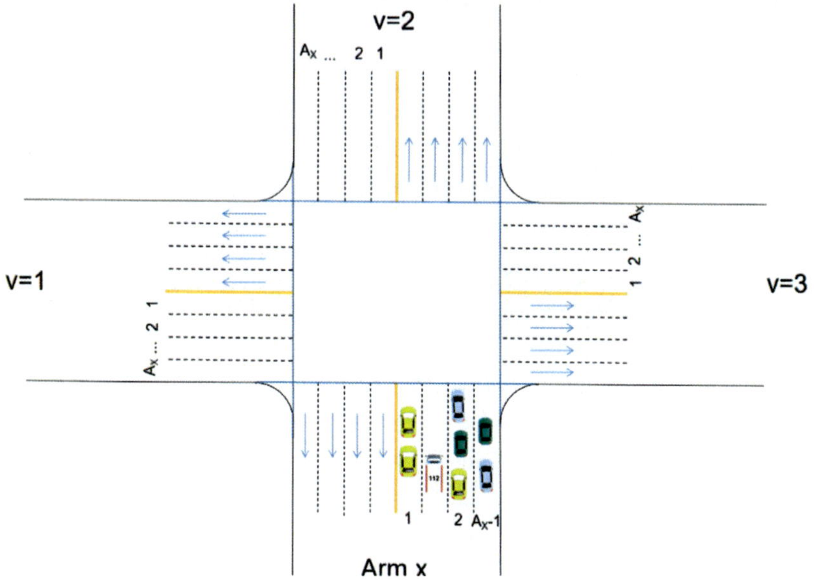

Fig. 2. Emergency vehicle occupancy

In view of the characteristics of the smart connected vehicles, the following assumptions are made.

1. All vehicles are intelligent connected vehicles.
2. All common vehicles voluntarily clear one lane for emergency vehicles through V2V and V2I communication.
3. We assume perfect communication.

4. Pedestrian-related factors (e.g., pedestrian-vehicle conflicts) are ignored to focus on the studied problem (Tables 1 and 2).

Table 1. The set of input parameters

Notations	Descriptions
T_s	Traffic arms
A_x	Approaching lanes
B_x	Exit lanes
m	From arm x to arm c through lane q
γ	A collection of incompatible movements
γ_s	A collection of incompatible signal groups
F_{m1m2}	An interval of incompatible movements
$D_{x,v,q}$	The binary variable determines the allowable movements at the lane
λ	Common flow multiplier
$O_{x,v}$	Traffic demand matrix at intersections
$O_{x,v,q}$	The assigned flow is distributed from arm x to arm c
$S_{x,q}$	Saturation flow of lane q in arm x
$r_{x,v,q}$	The radius of the turning trajectory
$f_{x,v,q}$	Flow ratio
$\overline{S}_{x,q}$	Saturated flow in a straight moving lane
$\Psi_{x,v}$	Green start time
$P_{x,v}$	Green duration
$\beta x,q$	The start of the green display received on lane q in arm x
$\Omega x,q$	The duration of the green display received on lane q in arm x
$g_{x,v}$	Minimum green duration
$N_{x,q}$	Maximum acceptable saturation of lane q in arm x
η	The total number of arms from x to c
M	Any large number
c	Cycle length
δ	Reciprocal of the cycle length
$\cap$	Successor function
e	The difference between effective green time and actual green time
x	Arm x can be numbered as 1, 2, 3, 4
v	Number clockwise with respect to x
q	Lane number

Table 2. The set of decision variables

Variables set		
Binary variables	$D_{x,v,q}$	The binary variable determines the allowable movements at the lane
	$\cap_{x,v,i,d}$	The order in which signal groups (x, v), (i, d) are displayed
Continuous variables	λ	Common flow multiplier
	$\Psi x,v$	Green start time
	$P_{x,v}$	Green duration
	$B_{x,q}$	The start of the green display received on lane q in arm x
	$\Omega x,q$	The duration of the green display received on lane q in arm x
	δ	Reciprocal of the cycle length
	$O_{x,v,q}$	The assigned flow is distributed from arm x to arm c

The developed model is as follows.

$$\text{Maximize } \lambda \tag{1}$$

Subject to

$$\sum\nolimits_{c=1}^{T_s-1} D_{x,v,q} \geq 1 \;\; \forall q = 1, \ldots, A_x; \quad \forall x = 1, \ldots, T_s \tag{2}$$

$$\begin{aligned} &\lambda O_{X,v} = \sum\nolimits_{q=1}^{A_x} O_{x,v,q} \quad \forall\, x = 1, \ldots, T_s; \quad \forall\, v = 1, \ldots, T_s - 1 \\ &MD_{x,v,q} \geq O_{x,v,q} \;\; \geq 0 \end{aligned} \tag{3}$$

$$\forall\, x = 1, \ldots, T_s; \quad \forall\, v = 1, \ldots, T_s - 1; \quad \forall\, q = 1, \ldots, A_x \tag{4}$$

$$1 \geq \psi_{x,v} \geq 0 \quad \forall\, x = 1, \ldots, T_s; \quad \forall\, v = 1, \ldots, T_s - 1 \tag{5}$$

$$1 \geq \rho_{x,v} \;\; \geq g_{x,v}\, \delta \quad \forall\, x = 1, \ldots, T_s; \quad \forall\, v = 1, \ldots, T_s - 1 \tag{6}$$

$$M\left(1 - D_{x,v,q}\right) \geq \beta_{x,q} - \psi_{x,v} \geq -M\left(1 - D_{x,v,q}\right) \tag{7}$$

$$M\left(1 - D_{x,v,q}\right) \geq \Omega_{x,q} - \rho_{x,v} \geq -M\left(1 - D_{x,v,q}\right) \tag{8}$$

$$\frac{1}{C_{min}} \geq \delta \geq \frac{1}{C_{max}} \tag{9}$$

$$\begin{aligned} &B_{\eta(x,v)} \geq \sum\nolimits_{q=1}^{A_x} D_{x,v,q} \quad \forall\, x = 1, \ldots, T_s; \quad \forall\, v = 1, \ldots, T_s - 1 \\ &1 - D_{x,v,q+1} \;\; \geq D_{x,d,q} \;\; \geq D_{x,v,q+1} \;\; \forall\, d = v + 1, \ldots, \; T_s - 1; \quad \forall\, v = 1, \ldots, T_s - 2 \end{aligned} \tag{10}$$

$$\begin{aligned} &\forall\, q = 1, \ldots, A_x - 1 \quad \forall\, x = 1, \ldots, T_s \\ &m_1 = (x, v, q) \quad m_2 = (i, d, t) \end{aligned} \tag{11}$$

$$\psi_{id} + \cap_{x,v,i,d} + M\left(2 - D_{x,v,q} - D_{i,d,t}\right) \geq \psi_{x,v} + \rho_{x,v} + F_{m1m2}\delta \quad \forall\, m_1, m_2 \in \gamma \tag{12}$$

$$\cap_{x,v,i,d} + \cap_{i,d,x,v} = 1 \quad \forall\, ((x, v), (i, d)) \in \gamma_s \tag{13}$$

$$\begin{aligned} &M\left(2 - D_{x,v,q} - D_{x,v,q+1}\right) \geq \frac{1}{Sx,q} \sum\nolimits_{v=1,\ldots,T_s-1} (1 + \frac{1.5}{r_{x,v,q}}) O_{x,v,q} \\ &- \frac{1}{Sx,q+1} \sum\nolimits_{v=1,\ldots,T_s-1} (1 + \frac{1.5}{r_{x,v,q}} O_{x,v,q+1}) \geq -M\left(2 - D_{x,v,q} - D_{x,v,q+1}\right) \\ &\forall\, q = 1, \ldots, A_x; \quad \forall\, x = 1, \ldots, T_s \end{aligned} \tag{14}$$

$$\begin{aligned} &\beta_{xq} + e\delta \geq \frac{1}{\overline{S}_{x,q} N_{x,q}} \sum\nolimits_{v=1,\ldots,T_s-1} ((1 + \frac{1.5}{r_{x,v,q}}) O_{x,v,q}) \\ &\forall\, q = 1, \ldots, A_x; \quad \forall\, x = 1, \ldots, T_s \end{aligned} \tag{15}$$

In the constraints listed above, Eq. 1 represents the objective of the optimization, where capacity is positively correlated with traffic demand, and the problem can thus be converted into solving the common flow multiplier. Equation 2 indicates that movement should be greater than 1 in each lane. Equation 3 ensures the normal operation of the intersection as traffic demand increases. Equation 4 defines if lane q indicates no movement, then there is no traffic allocation in the lane. Equation 5 and Eq. 6 define the constraints on green start time $\psi_{x,v}$ and $P_{x,v}$. Equation 7 and Eq. 8 ensure that all traffic movements on a lane q are consistent, with the green starting at the same time as the green received on the lane, and green duration time equals the green on the lane. Equation 9 defines the range of cycle length, ensuring that c is a reasonable value. Equation 10 indicates that the number of exit lanes on lane v should be greater than or equal to the number of lanes x in one movement.

Equation 11 denotes the lane q on the arm x is prohibited from turning onto other arms which do not include arm v when the lane q + 1 on the arm x allows turning onto the arm v to avoid unnecessary traffic collisions between adjacent lanes. Equation 12 defines the movements (x, v, q) (i, d, t) that a set of conflicts γ, and when both movements are allowed, we set the clearing time to avoid their conflicts. Equation 13 indicates the sequence of conflicting signaling groups. Equation 14 restricts the flow factor between adjacent lanes to make them equal. Equation 15 defines that the saturation on the lane q cannot exceed its maximum saturation $N_{x,q}$.

3 Case Study

In this section, we demonstrate the impacts of the EVs on the normal traffic flow at intersections with a series of numerical experiments. In order to solve the model in this paper, we apply it to the four-branch road intersection. By constraint (2)–(15) and mixed integer programming algorithm, a mixed integer linear programming model is established.Traffic signal optimization before and after the EV arrivals are conducted with the commercial solver Gurobi, respectively. At the studied four-leg intersection, the number of arms is set at 4. Since emergency vehicle traffic occupies one lane, one of the four legs can be reconsidered with 3 approaching lanes. Considering the principle that the number of exit lanes is greater than the number of approaching lanes, the exit lanes of each arm remain 4. The maximum cycle length is set to be 120 s. Assume that the minimum duration of all vehicle movements equals 5 s with a turning radius of 12 m. The maximum acceptable saturation per lane is set at 0.9. The time difference between a valid green and an actual green is set as 1 s, and the clear time for movements with conflicting trajectories is set as 3 s (Table 3).

Table 3. Pre-set values of parameters

Parameter	Interpretation	Value
T_s	Traffic arm	4
A_x	The approaching lane where arm X is located	4 or 3
B_η	The exit lanes where arm η is located	4
C_{max}	Maximum cycle length	120
$g_{x,v}$	Minimum green duration	5
$r_{x,v,q}$	Radius of the turn trajectory	12
$N_{x,q}$	Maximum acceptable saturation of lane q in arm x	0.9
e	The difference between effective green time and actual green time	1
F_{m1m2}	An interval of incompatible movements	3

To demonstrate the results of allocating one exclusive lane for the EV. All simulation scenarios are implemented via the python platform. Specifically, before the arrival of EV, we assume λ = 2.23 c = 120 s, while after the EV arrival, we assume λ = 2.26 c = 120 s.

The results of optimization indicate the maximum capacity of the direction of motion based on the pre-set saturation flow. X is calculated using the ratio of the total assigned flow in the direction of motion to the capacity of the traffic, and then the delay is calculated according to Eq. 16 (Fig. 3 and Table 4).

$$d = \frac{C\left(1-\frac{g}{C}\right)^2}{2(1-Xg/C)} \tag{16}$$

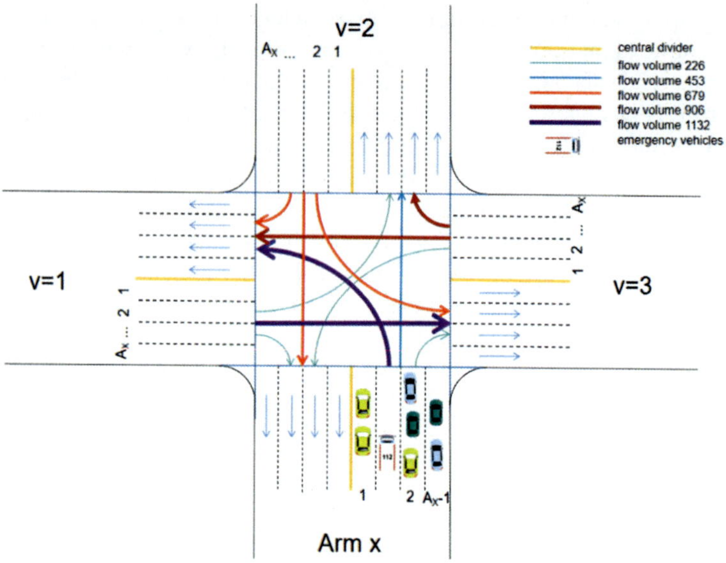

Fig. 3. Flow distribution at intersections

Table 4. Optimal results for one lane occupied by emergency vehicles

From x	To v	Total assigned flow (pcu/h)	Delay (s)	Traffic capacity (pcu/h)	$P_{x,v}$
1	1	1132	38.53	1227	0.33
1	2	452	29.94	739	0.38
1	3	226	25.97	739	0.38
2	1	226	52.05	236	0.13
2	2	1132	38.62	1227	0.33
2	3	226	52.26	236	0.12
3	1	679	34.10	739	0.10
3	2	679	40.23	962	0.26
3	3	679	40.01	986	0.26
4	1	226	52.07	236	0.13
4	2	906	43.22	975	0.26
4	3	906	24.96	991	0.54

In Fig. 4 and Fig. 5, group 1 represents the cases of lane occupancy by emergency vehicles and group 2 indicates the scenario before EV arrival. As can be seen from Fig. 4, delays generally increased due to the EV. Calculations of the overall average delay show that the overall average delay increased by a modest 11.28%. The movement to the right of the north entrance observed the biggest increase in delays at 36.63%. Of these, delays decreased in four directions, with the largest decrease of 30.94%. Comparisons of the traffic capacity of the two optimization results can be seen in Fig. 5. Following the occupancy of lanes by emergency vehicles, the overall average capacity decreased by 6.63%.

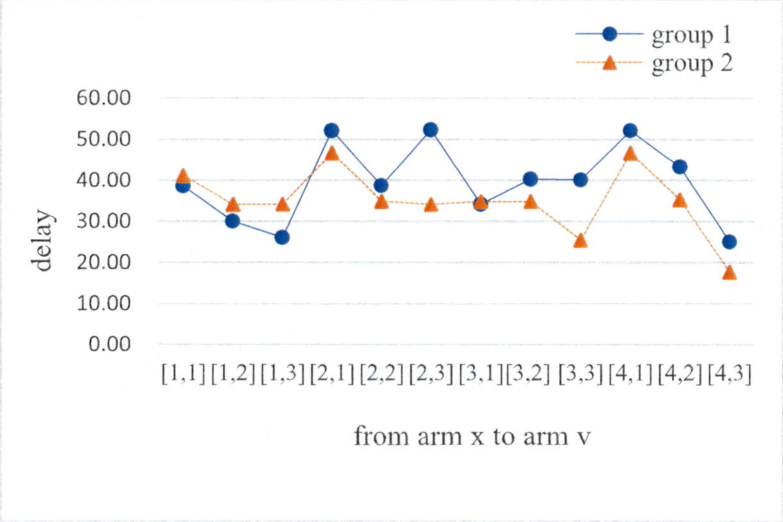

Fig. 4. Delay comparison between two optimizations

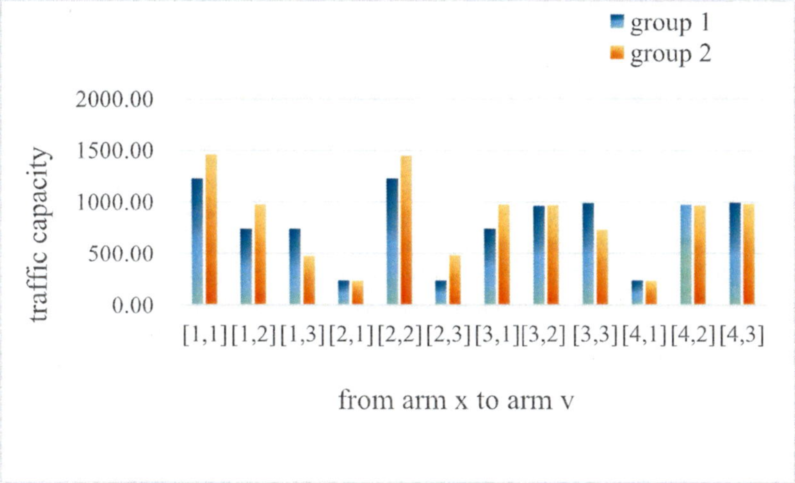

Fig. 5. Traffic capacity comparison between two optimizations

4 Conclusion

In this paper, we developed an optimization framework to evaluate the impact of aggressive emergency vehicles preemption strategies under optimal intersection control. A mixed-integer linear programming model is developed and solved by the Gurobi software. Numerical simulation results indicate that delay and capacity changes vary, with the overall average delay increase by 11.28% and the overall average capacity decrease by 6.63%. Future extension of the present study can be exploring the possibility of incorporating EV route choice with intersection optimization.

References

1. Calixto, E., Larouvere, E.: The regional emergency plan requirement: application of the best practices to the Brazilian case. Saf. Sci. **48**, 991–999 (2010)
2. Shibuya, S., Yoshida, T., Yamashiro, Z., Miyawaki, M.: Fast Emergency Vehicle Preemption Systems. Transp. Res. Rec. J. Transp. Res. Board. **1739**, 44–50 (2000)
3. Huang, Y., Weng, Y., Zhou, M.: Design of traffic safety control systems for emergency vehicle preemption using timed petri nets. IEEE Trans. Intell. Transp. Syst. **16**, 2113–2120 (2015)
4. Salih, A., Lewis, A., Chung, E.: Dynamic preemption algorithm to assign priority for emergency vehicle in crossing signalised intersection. IOP Conf. Ser. Mater. Sci. Eng. **518**, 022038 (2019)
5. Lee, J., Park, B.: Development and evaluation of a cooperative vehicle intersection control algorithm under the connected vehicles environment. IEEE Trans. Intell. Transp. Syst. **13**, 81–90 (2012)
6. Lu, C., Liu, C.: Ecological control strategy for cooperative autonomous vehicle in mixed traffic considering linear stability. J. Intell. Connect. Veh. **4**, 115–124 (2021)
7. Xu, Y., Ye, Z., Wang, C.: Modeling commercial vehicle drivers' acceptance of advanced driving assistance system (ADAS). J. Intell. Connect. Veh. **4**, 125–135 (2021)
8. Wu, J., Kulcsár, B., Ahn, S., Qu, X.: Emergency vehicle lane pre-clearing: from microscopic cooperation to routing decision making. Transp. Res. Part B Methodolog. **141**, 223–239 (2020)
9. Peng, B., Keskin, M., Kulcsár, B., Wymeersch, H.: Connected autonomous vehicles for improving mixed traffic efficiency in unsignalized intersections with deep reinforcement learning. Commun. Transp. Res. **1**, 100017 (2021)
10. Wu, J., Ahn, S., Zhou, Y., Liu, P., Qu, X.: The cooperative sorting strategy for connected and automated vehicle platoons. Transp. Res. Part C Emerging Technol. **123**, 102986 (2021)
11. Ali, Y., Zheng, Z., Haque, M.: Modelling lane-changing execution behaviour in a connected environment: a grouped random parameters with heterogeneity-in-means approach. Commun. Transp. Res. **1**, 100009 (2021)
12. Zhu, W., Wu, J., Fu, T., Wang, J., Zhang, J., Shangguan, Q.: Dynamic prediction of traffic incident duration on urban expressways: a deep learning approach based on LSTM and MLP. J. Intell. Connect. Veh. **4**, 80–91 (2021)
13. Kong, X., Wu, J., Qu, X.: An online processing method for the cooperative control of connected and automated vehicle platoons. In: Smart Innovation, Systems and Technologies, pp. 133–139 (2021)

Spatiotemporal Distribution of Traffic Violations in a Medium-Sized City Luzhou

Haiyue Liu[1], Yue Zhou[1], Chuanyun Fu[2](✉), and Yining Tan[1]

[1] School of Transportation and Logistics, Southwest Jiaotong University, Chengdu, China
[2] School of Transportation Science and Engineering, Harbin Institute of Technology, Harbin, China
fuchuanyun@hit.edu.cn

Abstract. Exploring the spatial and temporal distribution of traffic violations is vital to road safety management. This study investigated the traffic violations of illegal parking and disobeying the guide lane, which are the most observed in a middle-sized city Luzhou. The temporal distributions between the traffic violations are compared in time of day, day of week, and month of year. The underlying spatial dependency and cluster of the violations are investigated by global Moran's *I*, local Moran's *I*, and kernel density. Results show that the frequency of illegal parking is remarkably higher in the period ranging from September to December, weekdays, morning, and afternoon. However, the violations of disobeying the guide lane are more frequently observed in the first-half year and daytime. Both two types of violation are positively correlated in space. Moreover, the density of illegal parking is higher in the upper area of the city where commercial and residential zone are common, while disobeying the guide lane is mostly found in several intersections which are close to the freeway exit. The possible explanations of the spatial and temporal distributions are discussed.

Keywords: Traffic violation · Spatiotemporal distribution · Moran's I test · Kernel density analysis · Illegal parking · Disobeying the guide lane

1 Introduction

Numerous studies have pinpointed that drivers who commit traffic violations frequently were more likely to be involved in crashes [1–3]. Accordingly, traffic violation is viewed as one of the leading causes of road trauma [4, 5]. However, Traffic violations are much more common in developing countries [1, 6]. With these facts in mind, it is worth preventing urban traffic violations with effective approaches.

Generally, the researchers and authorities of road safety tend to focus on several types of urban traffic violations, which are regarded as risky behaviors, such as speeding [7–10], driving under the influence [11–15]. In contrast to those, others including illegal overtaking [16, 17], disobeying the guide lane [18] (driver should use guide lanes to enter intersections to avoid potential conflicts), and illegal parking [19] are less concerned, although they are sometimes more common in the city areas than the "risky violations".

Y. Bie et al. (Eds.): KES-STS 2022, SIST 304, pp. 161–170, 2022.
https://doi.org/10.1007/978-981-19-2813-0_17

The guide lane is a turning lane generally set in the entering section of the intersection. Drivers whose turning are inconsistent with the guide lane it is in will get tickets for disobeying the guide lane. Vehicles are not allowed to park in front of the entering of public service buildings, intersections, and narrow roads. Such a violation is called illegal parking. It should be noted that disobeying the guide lane and illegal parking (especially on-road parking) can affect the traffic flow by causing hard-braking and traffic conflict [20–22], which is considered to be a pre-crash improper behavior [23] and significantly associated with pedestrian-related crashes [24]. As such, it is urgent to explore the characteristics of such two types of traffic violations.

The spatial and temporal distribution of traffic violations is commonly investigated. Red-light running was inclined to be detected on alleys, roads with physical dividers, and intersections [25] on weekends, midday, morning peak hours, and evening [26], speeding behaviors are more likely to be found at nighttime [27], and driving without seat belt is more commonly observed on urban roads rather than rural roads [28, 29]. Moreover, traffic violations are found to be associated with both temporal and spatial autocorrelations [30], with certain clustering patterns in space and time are observed. Traffic engineers can have a better understanding of determinants spatiotemporal characteristics and autocorrelation effects in traffic violations. However, the studies have several limitations. In addition to the incomplete researches on some types of violations, the existing study often uses the data of traffic violations collected from the whole country, state, and metropolitan, while the violations from minor sizes (such as medium-sized cities) are ignored.

To fill the knowledge gaps in the literature, this study attempts to explore both the spatial and temporal distribution characteristics of two types of traffic violations, i.e., disobeying the guide lane and illegal parking. Luzhou, a medium-sized city located in Southwestern China, is selected to be analyzed. The temporal and spatial patterns of the traffic violations are investigated on different scales. This effort could provide guidance for road safety management in similar cities.

2 Method

2.1 Data

Data of traffic violations were collected from the Department of Policy in Luzhou City, 2016. Luzhou is a medium-sized city in Sichuan Province, China. The core urban area of this city consists of three districts, namely Jiangyang, Longmatan, and Naxi. Both Jiangyang and Longmatan districts are located in the upper area of the city, while the Naxi district belongs to the lower area of the city. Luzhou is a typical industrial and port city in mainland China, which has considerable needs for cargo shipping and a high ratio of citizens engaged in logistics.

According to data processing, 14 types of traffic violations are identified and ranked in frequency, as shown in Fig. 1. It shows that the violations of illegal parking and disobeying the guide lane make up the majority of total traffic violations in Luzhou (84.6%), while other violations only account for minor parts. Due to the dominance of illegal parking and disobeying the guide lane, the two types of traffic violations are selected to be investigated.

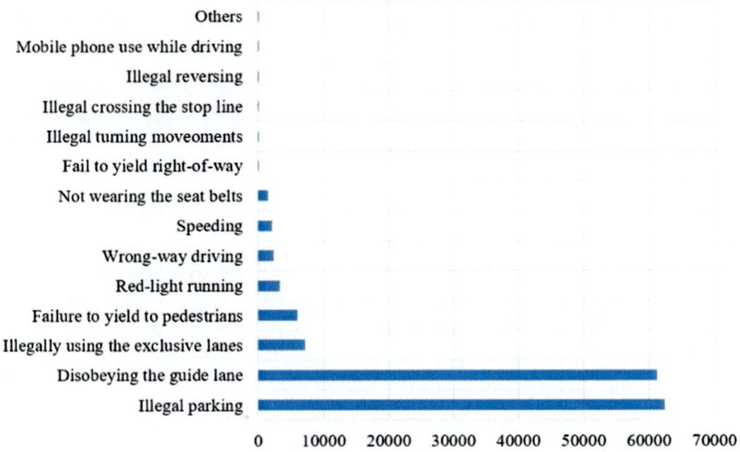

Fig. 1. Frequency of 14 types of traffic violation in 2016

2.2 Analytical Approach

The approaches used to analyze the spatial and temporal distributions of the traffic violations are presented as follows.

1) **Temporal analysis**

Temporal distributions of the traffic violations are measured by descriptive statistics. We explore the distribution of the violation frequency in three-time scales, including time-of-day, day of week, and month of year.

2) **Spatial analysis**

We use global Moran's *I* test, local Moran's *I* test, and kernel density analysis to unveil the spatial pattern of the violations. Global Moran's *I* and local Moran's *I* require spatial units with attributes of interest (e.g., points, lines, grids). Thus, the study area is divided into grids with violation frequency to be analyzed.

(1) Global Moran's *I* test

Global Moran's *I* test is used to examine the underlying spatial autocorrelation of a variable in the entire study area. It can be given by

$$Moran's\, I = \frac{N}{\sum_{ij} w_{ij}} \frac{\sum_{i}\sum_{j} w_{ij}(x_i - \overline{x})(x_j - \overline{x})}{\sum_{i} (x_i - \overline{x})^2} \tag{1}$$

where, x_i and x_j are the frequency of a type of traffic violation in the grid i and j, respectively; $\overline{x}$ is the mean value of the traffic violation; N is the total number of the grids; w_{ij} is the spatial weight, specifically, if grid x_i and grid x_j are adjacent, then $w_{ij} = 1$, otherwise, $w_{ij} = 0$.

The interval of global Moran's I statistic is $[-1, 1]$. If the statistic is close to -1, the attributes of interest are more negatively correlated in space. If the statistic is close to 1, the attributes are more positively correlated in space.

(2) Lobal Moran's I test

Moreover, the study uses a local Moran's I test to explore whether a local spatial autocorrelation existing among the adjacent grids [31], it takes the form as

$$I_i = \frac{x_i - \overline{x}}{S^2} \sum_{j}^{n} w_{ij}(x_j - \overline{x}) \tag{2}$$

where S^2 is the variance of x_j.

Local Moran's I test sorts the grids into different patterns of spatial clusters, which are high-high cluster, high-low cluster, low-high cluster, and low-low cluster. High-high clusters and low-low clusters mean that the grids with high violation frequency or low violation frequency are surrounded by similar grids, respectively; while the high-low or low-high clusters mean the target grid which has a high/low value of the violation frequency is surrounded by the dissimilar grids.

(3) Kernel density analysis

In addition to the grid-based approaches, we use kernel density analysis to reflect the continuous spatial distribution of the traffic violations on point elements. Generally, geographic events are more likely to occur in areas with high kernel density and less likely to occur in areas with low kernel density. The equation of kernel density is expressed as

$$f_n(z) = \frac{1}{nh} \sum_{i=1}^{n} k(\frac{z - z_i}{h}) \tag{3}$$

where, $k(\cdot)$ is the kernel function; h is the bandwidth; z represents the points with attributes; z_i is the adjacent points of z.

3 Results

3.1 Temporal Characteristics

Figure 2 shows the temporal distribution of the traffic violations in time-of-day, which are presented by bar charts. It shows that the violations of illegal parking are mostly found in the morning (9:30–11:30) and afternoon (15:30–16:30) (Fig. 2 (a)). A slight incensement is also found after the evening peak hours (19:30–20:30). For the number of disobeying the guide lane, it is commonly found during the whole daytime (8:30–19:00). We also observe an extreme of such violation in the evening peak hours (see Fig. 2 (b)).

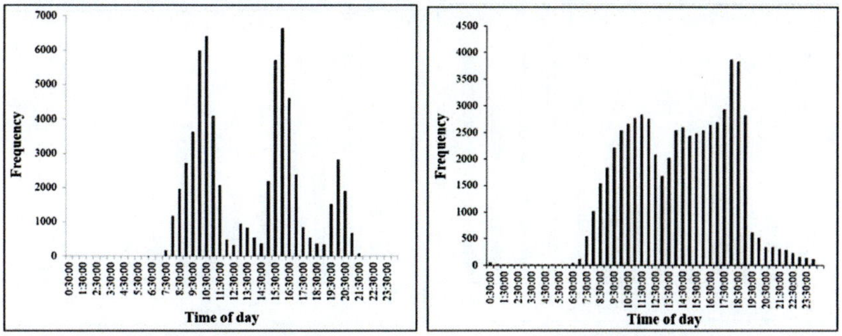

(a) Illegal parking (b) Disobeying the guide lane

Fig. 2. The time-of-day distribution of the two types of traffic violation

Neither the two types of traffic violation are inclined to be found in late night and dawn (21:30–6:30).

Figure 3 illustrates the distribution on the day of week for the illegal parking and disobeying the guide lane. It can be seen that Monday and Tuesday are the two days with the highest frequencies of illegal parking (12,240 and 11,295 tickets), while Saturday and Sunday have lower frequencies of illegal parking than other days (see Fig. 3 (a)). As to the distribution of disobeying the guide lane, it seems no obvious difference is observed between weekdays and weekends, although the frequency of Monday and Friday (10030 and 9618 tickets) slightly surpasses this of other days.

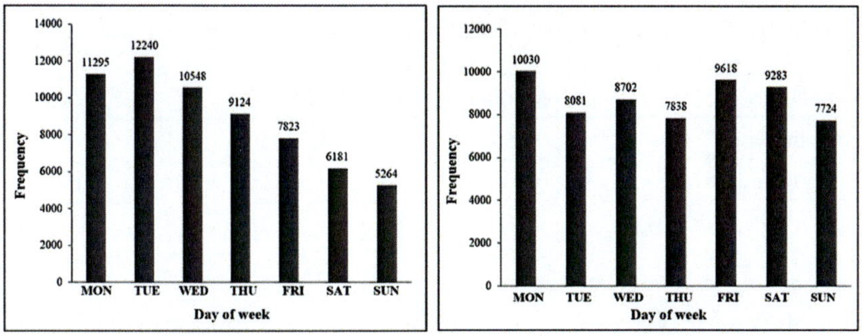

(a) Illegal parking (b) Disobeying the guide lane

Fig. 3. The day of week distribution of the two types of traffic violation

The monthly distribution of the two types of traffic violations is illustrated in Fig. 4. Results show that illegal parking is more likely to occur in the range from September to December (Fig. 4 (a)), yet the frequency of these violations are much fewer in the first half-year. However, the monthly trend of disobeying the guide lane is quite discrepant in contrast to illegal parking. These violations are commonly observed in the first half-year with the extreme in January, while the frequency at the end of this year is quite low (see Fig. 4 (b)).

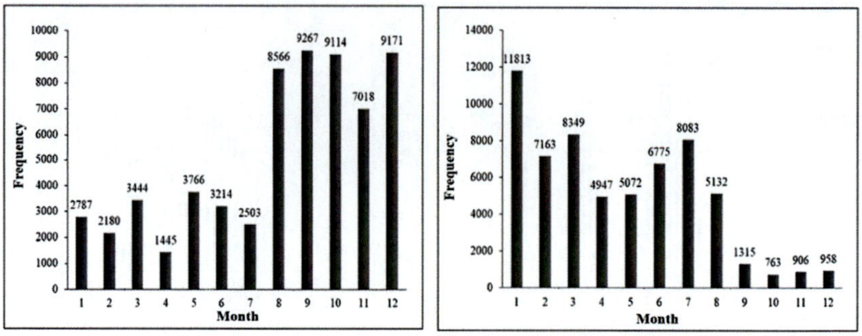

(a) Illegal parking (b) Disobeying the guide lane

Fig. 4. The monthly distribution of two types of traffic violations

3.2 Spatial Characteristics

1) Global Moran's *I* Test

As presented in Table 1, the pseudo *p*-values of the global Moran's *I* statistic show that both the violations of illegal parking and disobeying the guide lane are not randomly distributed in the study area, indicating that there are significant spatial autocorrelations among these violations. Moreover, we found the global Moran's *I* statistic is positive for either the types of violation. This signifies that the violations are spatially clustered rather than dispersed.

Table 1. The results of global Moran's *I* test of two types of traffic violations

Traffic violations	Moran *I*	*Z*-value	*p*-value
Illegal parking	0.2687	311.3763	$p < 0.01$
Disobeying the guide lane	0.0013	2.3451	$p < 0.05$

2) Local Moran's *I* Test

Results of the local Moran's *I* test are shown in Fig. 5. Grids with high-high clusters are mostly found in the center and the southwest corner of the city. This presents that the grids with the high frequency of illegal parking tend to be correlated. Only one grid of low-high outliers is observed in the city center, indicating that the frequency of illegal parking in this grid is relatively less than its neighbors. With respect to the violation of disobeying the guide lane, the grids with high-high clusters are much fewer than those of illegal parking in the upper area of the city. It is also found that a grid with a high-low cluster is located downstream of urban rivers, meaning that the grid with the high frequency of disobeying the guide lane is surrounded by the grids with less violation frequency.

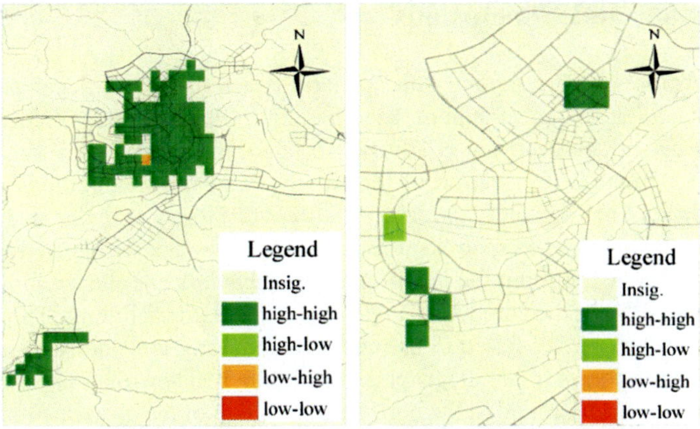

(a) Illegal parking **(b) Disobeying the guide lane**

Fig. 5. The results of local autocorrelation analysis of the two types of traffic violation

3) Kernel Density

The results of kernel density are presented in Fig. 6 (a) and Fig. 6 (b). It is shown that the density of illegal parking in the upper area surpasses that of other areas where land-use is typed as a commercial and residential zone. Moreover, the location with public and residential land-use tends to have a higher density of illegal parking as well. The kernel density analysis for disobeying the guide lane shows that only one location is observed with extremely high density, which is located in the northern part of the city and close to the freeway exit. Some public facilities and commercial buildings are also found in this location, including hospitals, pharmacies, and guesthouse complexes.

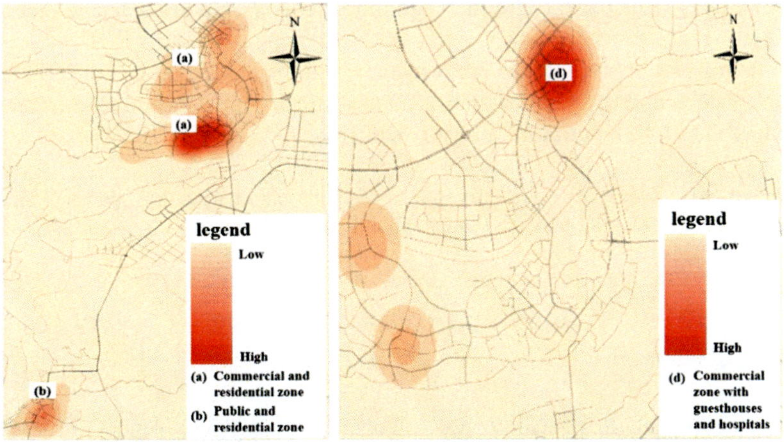

(a) Illegal parking **(b) Disobeying the guide lane**

Fig. 6. The result of kernel density of the two types of traffic violations

4 Discussions and Conclusions

This study explores both temporal and spatial characteristics of illegal parking and disobeying the guide lane in a medium-sized port city. Global Moran's I, local Moran's I, and kernel density analysis are conducted to uncover the spatial patterns. We also investigate the temporal distribution of the two types of traffic violations on various scales (hourly, day of the week, and monthly). Results show that the frequencies of traffic violations are linked to spatial autocorrelation and spatial-temporal variation.

For the spatial distribution, the illegal parking behaviors are clustered in uncertain areas. The areas serve as the core with a huge amount of traffic activities. The increasing requirements of parking are failed to be met by the parking lots and roadside parking spaces, resulting in considerable illegal parking violations. The violations of disobeying the guide lane are mostly found in the northeastern region of the city. One reason is that the intersections, which are installed with cameras, are located in an arterial close to the freeway exit. Thus, drivers who are unfamiliar with the lane-setting (e.g., cargo shipping drivers and travelers from other cities) may commit the violation easily. Another reason may be attributed to the setting of the camera. Compared to other cameras in other regions, those installed in this region are covert, which could hamper the alert of the surrounding drivers.

For the temporal distribution, the illegal parking violations are mostly detected in the morning, afternoon (before the evening peak hours), and working days. This could be that most vehicles are temporarily parking on the roadside due to business reasons or dining needs. Nevertheless, the frequencies of disobeying the guide lane are uniform during the same range as the traffic volume, and underlying offenders passing the detection point are very large. Interestingly, the two types of traffic violations are remarkably discrepant across the months. It could be that the inner-city traffic activities are less intensive in the first half-year due to the imbalanced activities of such a port city. As a result, it leads to fewer illegal parking violations. Conversely, the city has more cargo-shipping needs from other cities in this period, and there are more vehicles that are detected nearby the freeway exit.

A limitation of our study is the analysis may be affected by the layout of detection devices. New detection approaches such as naturalistic driving data and GPS trajectories are encouraged to figure out more robust conclusions [32–34].

Acknowledgments. This study was jointly sponsored by the National Natural Science Foundation of China (Grant No. 71801182) and the Fundamental Research Funds for the Central Universities (Grant No. FRFCU5710000111).

References

1. Shaaban, K., Abdalla, S., Badran, A.: Analysis of traffic crashes and violations in a developing country. Transp. Res. Procedia **55**, 1689–1695 (2021)
2. Lourens, P., Vissers, J., Jessurun, M.: Annual mileage, driving violations, and accident involvement in relation to drivers' sex, age, and level of education. Accid. Anal. Prev. **31**(5), 593–597 (1999)

3. Yoh, K., Okamoto, T., Inoi, H., Doi, K.: Comparative study on foreign drivers' characteristics using traffic violation and accident statistics in Japan. IATSS Res. **41**(2), 94–105 (2017)
4. Fu, C., Sayed, T.: Multivariate Bayesian hierarchical Gaussian copula modeling of the non-stationary traffic conflict extremes for crash estimation. Anal. Methods Accid. Res. **29**, 100154 (2021)
5. Fu, C., Sayed, T., Zheng, L.: Multivariate Bayesian hierarchical modeling of the non-stationary traffic conflict extremes for crash estimation. Anal. Methods Accid. Res. **28**, 100135 (2020)
6. Road Traffic Safety Research Centre of the Ministry of Public Security: Research report of metropolis' road traffic development in China. Road Traffic Safety Research Centre of the Ministry of Public Security, Beijing (2018)
7. Truelove, V., Freeman, J., Kaye, S., Watson, B., Mills, L., Davey, J.: A unified deterrence-based model of legal and non-legal factors that influence young driver speeding behaviour. Accid. Anal. Prev. **160**, 106327 (2021)
8. Tucker, A., Marsh, K.: Speeding through the pandemic: perceptual and psychological factors associated with speeding during the COVID-19 stay-at-home period. Accid. Anal. Prev. **159**, 106225 (2021)
9. Ambo, T., Ma, J., Fu, C.: Investigating influence factors of traffic violation using multinomial logit method. Int. J. Inj. Contr. Saf. Promot. **28**(1), 78–85 (2021)
10. Zhou, Y., Jiang, X., Fu, C., Liu, H.: Operational factor analysis of the aggressive taxi speeders using random parameters Bayesian LASSO modeling approach. Accid. Anal. Prev. **157**, 106183 (2021)
11. Messmann, U., Torfs, K., Cools, M.: Socio-cognitive factors in road safety monitoring – cross-national comparison of driving under the influence of alcohol, drugs or medication. IATSS Res. **44**(3), 180–187 (2020)
12. Yadav, A.K., Velaga, N.R.: An investigation on the risk factors associated with driving errors under the influence of alcohol using structural equation modeling. Traffic Injury Prev. **21**(4), 288–294 (2020)
13. Smailovic, E., Lipovac, K., Pesic, D., Antic, B.: Factors associated with driving under the influence of alcohol. Traffic Inj. Prev. **20**(4), 343–347 (2019)
14. Hasan, R., Watson, B., Haworth, N., Oviedo-Trespalacios, O.: A systematic review of factors associated with illegal drug driving. Accid. Anal. Prev. **168**, 106574 (2022)
15. Mills, L., Watson-Brown, N., Freeman, J., Truelove, V., Davey, J.: An exploratory investigation into the self-regulatory processes influencing drug driving: are young drivers more externally regulated? Transp. Res. F Traffic Psychol. Behav. **80**, 237–249 (2021)
16. Huemer, A.H., Strauß, F.M.: Attitude vs. infrastructure: influences on the intention to overtake bicycle riders. Transp. Res. Interdiscip. Perspect. **10**, 100397 (2021)
17. Liu, Q., Sun, J., Tian, Y., Xiong, L.: Modeling and simulation of overtaking events by heterogeneous non-motorized vehicles on shared roadway segments. Simul. Model. Pract. Theory **103**, 102072 (2020)
18. Ayele, A.E., Jiang, X., Fu, C.: Spatial point pattern analysis of traffic violations in Luzhou City. China. Transp. Lett. **13**(10), 1–10 (2021)
19. Wang, S., Liu, Y.: Parking in inner versus outer city spaces: spatiotemporal patterns of parking problems and their associations with built environment features in Brisbane, Australia. J. Transp. Geogr. **98**, 103261 (2022)
20. Chen, T., Sze, N., Chen, S., Labi, S., Zeng, Q.: Analysing the main and interaction effects of commercial vehicle mix and roadway attributes on crash rates using a Bayesian random-parameter Tobit model. Accid. Anal. Prev. **154**, 106089 (2021)
21. Fu, C., Sayed, T., Zheng, L.: Multi-type Bayesian hierarchical modeling of traffic conflict extremes for crash estimation. Accid. Anal. Prev. **160**, 106309 (2021)
22. Fu, C., Sayed, T.: Random parameters Bayesian hierarchical modeling of traffic conflict extremes for crash estimation. Accid. Anal. Prev. **157**, 106159 (2021)

23. Wang, Z., Huang, S., Wang, J., Sulaj, D., Hao, W., Kuang, A.: Risk factors affecting crash injury severity for different groups of e-bike riders: a classification tree-based logistic regression model. J. Saf. Res. **76**, 176–183 (2021)
24. Yoon, J., Lee, S.: Spatio-temporal patterns in pedestrian crashes and their determining factors: application of a space-time cube analysis model. Accid. Anal. Prev. **161**, 106291 (2021)
25. Şimşekoğlu, Ö., Lajunen, T.: Relationship of seat belt use to health and driver behaviors. Transp. Res. F Traffic Psychol. Behav. **12**(3), 235–241 (2009)
26. Li, Y., Li, M., Yuan, J., Lu, J., Abdel-Aty, M.: Analysis and prediction of intersection traffic violations using automated enforcement system data. Accid. Anal. Prev. **162**, 106422 (2021)
27. Zhou, Y., Fu, C., Jiang, X., Mao, C., Liu, H.: Road factor analysis of taxi speeding behavior considering spatial effect. China Saf. Sci. J. **31**(3), 162–170 (2021)
28. Alver, Y., Demirel, M., Mutlu, M.: Interaction between socio-demographic characteristics: traffic rule violations and traffic crash history for young drivers. Accid. Anal. Prev. **72**, 95–104 (2014)
29. Zhang, G., Tan, Y., Jou, R.: Factors influencing traffic signal violations by car drivers, cyclists, and pedestrians: a case study from Guangdong, China. Transp. Res. F Traffic Psychol. Behav. **42**, 205–216 (2016)
30. Wang, S., Chen, Y., Huang, J., Liu, Z., Li, Z., Ma, J.: Spatial relationships between alcohol outlet densities and drunk driving crashes: an empirical study of Tianjin in China. J. Safety Res. **74**, 17–25 (2020)
31. Fu, C., Zhou, Y., Xu, C., Cui, H.: Spatial analysis of taxi speeding event using GPS trajectory data. In: Proceedings of the 2019 IEEE Intelligent Transportation Systems Conference, pp. 122–127. IEEE, New York (2019)
32. Liu, H., Fu, C., Jiang, C., Zhou, Y., Mao, C., Zhang, J.: Bayesian hierarchical spatial count modeling of taxi speeding events based on GPS trajectory data. PLoS ONE **15**(11), e0241860 (2020)
33. Chevalier, A., et al.: Naturalistic speeding data: drivers aged 75 years and older. Data Brief **8**, 136–141 (2016)
34. Fu, C., Sayed, T.: Bayesian dynamic extreme value modeling for conflict-based real-time safety analysis. Anal. Methods Accid. Res. **34**, 100204 (2022)

Scenario-Oriented Contract Based Design for Safety of Autonomous Vehicles

Nadra Tabassam(✉) and Martin Georg Fränzle(✉)

Carl von Ossietzky University Oldenburg, Oldenburg, Germany
{nadra.tabassam,martin.fraenzle}@uni-oldenburg.de

Abstract. The development and advances in the domain of Autonomous Vehicles (AVs) provide disparate benefits including improved safety, enhanced reliability, and reduction in accidents (which saves thousands of human lives). However, the guarantee of safety is crucial for the successful deployment of AVs on the road. The safety-related issues of AVs are disparate in nature and are not solvable with a single technological solution, but require interaction and distributed responsibility across numerous interactive components, rendering the evaluation of safety concerns a challenge. We have presented an Assume/Guarantee (A/G) based contract for solving the safety-related concerns of AVs. Our preeminent focus is on the collision avoidance of AVs. Our approach is based on 1. the formal specification of A/G based contracts 2. creation of collision-based scenarios, 3. implementation of scenario-oriented controller implementing A/G contracts, 4. testing of controller implementing A/G contracts (allows the verification of contracts in terms of minimizing collision risks) based on the simulation performed in CARLA Scenic while considering the different (collision oriented) scenarios. Each contract is based on the extended assumption considering a specific scenario, referred as scenario-oriented contracts. The proposed methodology shows adequate results and proves that the Contract-Based Design (CBD) can provide a propitious road map for solving the safety-related concerns of AV.

Keywords: Autonomous vehicles · Safety · CARLA · Scenic · Contract-based design

1 Introduction

The domains of mobility and transportation is changed profoundly due to expected advancements in autonomous driving. The computing power and functional complexity of Autonomous Driving Systems (ADSs) have increased adequately over the past few years. Factually, we only have prototypes of ADSs, and the safety of AV is preponderant for its safe deployment. ADSs are classified as safety-critical systems due to the societal dependence on their reliability and safety. The ever-growing heterogeneity and sophistication of these systems demand radical changes in the way safety-related requirements are modeled and

Y. Bie et al. (Eds.): KES-STS 2022, SIST 304, pp. 171–183, 2022.
https://doi.org/10.1007/978-981-19-2813-0_18

articulated. It seems imperative to enforce an extensive safety formalism during the design process of these systems, foreseeing and preventing catastrophic situations during operation. The performance and safety of AV (along with their successful launch) are highly dependent on the 1. driving policies adopted in different traffic scenarios (keeping in context the safety concerns) and 2. the huge amount of socio-temporal data extracted from the environment. Therefore, it is imperative to analyze the data (extracted during driving of AVs) in a manner that safe policies can be adopted for the minimizing the ratio of accidents. Therefore, it can be claimed that with driving policies based on rigorous coverage of 1. the behavior of the environment (assumptions pertain to the environment) and 2. the requirements the system fulfills relative to these (guarantees), the safety of autonomous driving can be justified.

Recent studies show that to ensure safety millions of miles of test driving are needed which require tens of years [1]. Formal verification methods complemented with simulation-based testing can fulfill the safety demands of AVs. Adopting CBD as a rigorous basis of these methods has recently gained momentum for the safety assessment of 1. Cyber-Physical Systems (CPS) [2] in general and 2. automotive prototypes [3]. Together with verification, it can improve the predictability and reliability of CPS [4]. The CBD provides a modular approach for system design and verification. A contract C is a pair of an assumption A and a guarantee G, represented as $C = (A, G)$. Assumptions refer to the characteristics or a set of behavioral preferences of the environment in which the design artifact operates while the guarantee explicates the artifact's behavior in environments conforming to the assumption [5].

Therefore, deriving the contracts with the set of assumptions (applicable to every possible safety situation that may occur during driving maneuvers) is a complex task. It is impossible to model all the differentiated scenarios in the contracts. But if we create the extended versions of contracts in complex scenarios then there is a possibility that the risk of catastrophic situation can be mitigated up to acceptable levels. Keeping this context in mind we leverage such assume-guarantee (A/G) based contracts (while defining some particular scenarios of collision in the assumptions) along with controller implementation and simulation-based testing for verifying the efficiency of these contracts. The controller implementation of these contracts is then programmed in CARLA Scenic so that evidence can be presented that the system design fulfills them (results presented in Sect. 5).

To our knowledge, this is the first paper in which safety-based contracts are created, implemented (through the controller), and tested (for finding the fact that they are fulfilling the specified requirements or not) relative to specific scenarios. Our major contribution is to prove that A/G-based contracts are capable of solving safety-related issues of AVs. We are trying to counter the collision-based challenges faced by AV by implementing scenario-oriented A/G contracts. The proposed methodology comprises the following steps:

1. Modelling of an environment (comprising of AVs, pedestrians, roads, and town) using CARLA Scenic
2. Creation of basic scenarios (Each scenario provides a situation in which the probability of collision between AVs is high). We have specified only three scenarios so that we can keep our assumptions simple and fewer. For any finite number of concrete scenarios, we can implement detailed mechanisms safeguarding exactly these scenarios without necessarily generalizing to other scenarios. Therefore, fewer assumptions are easier to satisfy due to the reduction of the criticality of requirements addressed by the proposed contract.
3. Creation of A/G based contract specifications (based on the scenarios specified in point 2) for avoiding collisions.
4. Creating controllers implementing contracts using CARLA Scenic.
5. Verifying the behavior of contracts in different scenarios for knowing the fact that they are capable of fulfilling the specified requirements or not?
6. Generating results 1. for analyzing the effect of creating safety controllers implementing contracts (for minimizing the risk of collision between AVs) and 2. to analyze that proposed contract-based specifications are capable of fulfilling safety-based requirements needed for the AVs.

The rest of the paper is structured a follows. Section 2 explains the existing work concerning contracts in AVs. Section 3 describes the proposed methodology including the creation of scenes, collision scenarios and contracts. Section 4 explains the CARLA simulation and code snippets for the implementation of contracts. Section 5 explains the results considering the two different cases. Section 6 describes the conclusion and future work.

2 Literature Review

Fremont et al. have tested the scenario-based safety of AVs using the simulation and real-world testing methodologies. They have answered the two questions in their research: 1. is simulation-based testing helpful for the safety-related concerns of AVs? and 2. is data collected from the simulation-based testing is accurate enough when compares to the data extracted from the real-world track testing (for avoiding AV accidents)? For building the test scenarios the programming language Scenic is used and for capturing the safety-related details of AV, metric temporal logic is employed. Therefore, a formal methods-based approach for evaluating the safety of machine learning AVs spanning safety properties and formal specification of scenarios is applied. Simulation-based testing along with the evaluation of the methodology (in both simulation and the real world) is performed [6]. Proposed methodology is closer to our work up to some extent in terms of selected scenarios and implementation of scenarios in Scenic. However, we have only selected three collision based scenarios and then employed CBD for avoiding collision.

Majzik et al. presents a system-level assurance technique for the black-box testing of AVs while employing the various model-based techniques. In order to extract qualitative data from the simulation runs, coherent and structured model queries are employed. Safety properties of AVs are continuously monitored (while creating different challenging scenarios) using graph queries and complex event processing techniques [7]. Authors in [8] have proposed the formal verification method for the normal behavior of AVs. Traffic situations and possible reactions of AVs are verified at the abstract level. Simulations are widely employed for analyzing the behavior of AVs [9]. However, the creation of diverse scenarios based on the collision along with the particular implementations satisfying the contract (for collision avoidance) and verification (that how these contract implementations can avoid collision in AVs) performed in simulation in the Scenic is not implemented before.

3 Methodology

This section describes the methodology followed for the implementation of our proposed technique. Figure 1 shows the steps followed in the proposed methodology.

Modelling of the Scene. The first step is the creation of a scene. A simulation scene SC_1 within the environment E is created. E comprises roads, lanes, trees, signals and AVs etc. For SC_1 an environment E is mathematically described as $E = (L, C_f, C_r)$. L represents the lane while C_f represents the front car and C_r represents the rear car. The creation of E is performed in Scenic which is domain-specific scenario description language and used for describing these types of environments [10]. The syntax of the language is designed in a way that it allows the user to create the disparate scenarios along with the implementation of soft and hard constraints. When executing Scenic, the CARLA server runs in the background. CARLA [11] is an open source driving simulator used for the purpose of autonomous driving research. CARLA provides the support for development, training and validation of AVs in the context of urban driving.

Creation of Scenario. In this step three different and basic scenarios S_1, S_2, S_3 are created within the SC_1 (created in previous step). Other scenarios S_4, S_5 and S_6 are the extensions of these basic scenarios. S_1, S_2 and S_3 are collision based scenarios. However, created scenarios may incur some increased risk of collision. At this point of time the SC_1 is already created and running in Scenic. All scenarios have two AVs C_f and C_r driving at different speeds (while speed of C_f is represented by V_f and speed of C_r is represented by V_r) with the safe distance (D) above 30 m. The three scenarios (represented by S_1, S_2, S_3) are based on the following situations (in which there is a high probability of collision due to the increase or decrease of speeds):

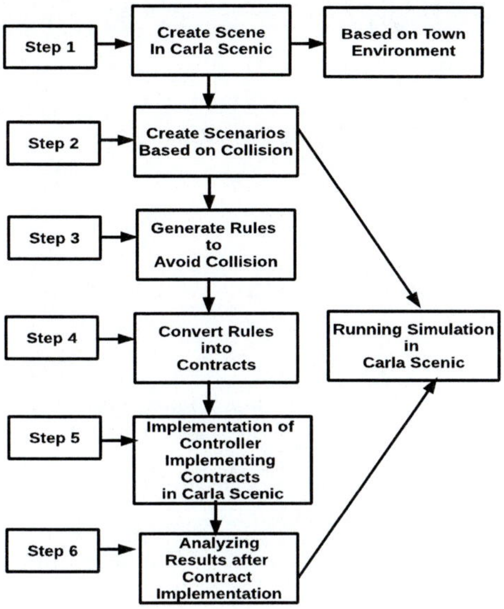

Fig. 1. Proposed methodology

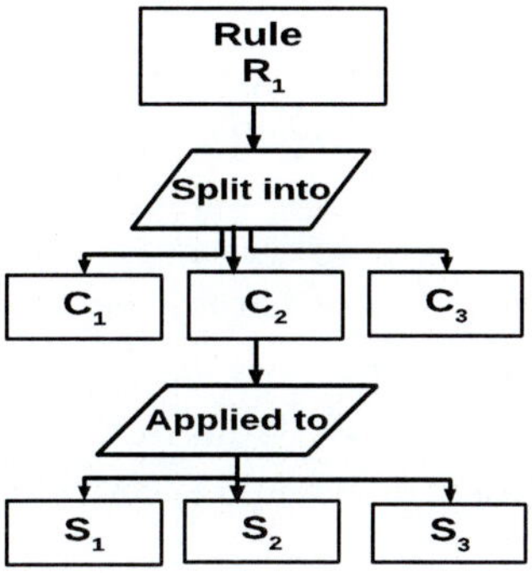

Fig. 2. Mapping of contracts on scenarios

S_1: After covering a distance of 5 km, C_r increases its speed and D between the C_f and C_r is less than 30 m.
S_2: After covering a distance of 5 km, C_f decreases its speed and D between C_f and C_r is now less than 30m at first and then D decreases to 10 m.
S_3: After covering a distance of 5 km, C_f decreases its speed and D between the C_f and C_r is less than 30 m. (The distance covered by the AVs (i.e.: 5 km) (after which we introduce the collision oriented situation is selected randomly)).

In order to keep the simulation simple only two AVs are considered within the SC_i rather than considering extremely busy road situations. However, making such assumptions explicitly opens up for automatic ways of detecting potential conflicts, thus rigorously uncovering potential incompleteness as well as potential conflicts in safety specifications.

Creation of Contracts. After the creation of collision based scenarios, the next step is based on the creation of contracts. Implementation of all contracts are performed in Scenic. The A/G contracts deal with collision avoidance (elaborated in S_1, S_2, S_3, S_4, S_5 and S_6). The contracts in our context are extracted from the mathematical rules defined in [8]. These rules have numerous dimensions, one of them is termed as "safe longitudinal distance-same direction". This rule is based on the fact that two AVs moving in the same direction should maintain a safe distance. The methodology for mapping contracts on the scenarios based on the rules is shown in Fig. 2. Whenever the contracts are satisfied in the scenarios and collision is avoided, these contracts (a program written for implementation of a contract monitor signalling satisfaction of the contract) return the true value for the variable Contract Implementation (CI). The rule (R_1) that is taken into consideration is defined as follows: R_1: Two AVs front AV (C_f) and rear AV (C_r) are driving in the same direction on the road with single lane.

Based on S_1: If the distance (D) between the C_f and C_r is reduced to less than 30 m as the C_r increases its speed then C_f should accelerate and increase its speed.

Based on S_2: In case if C_f decreases its speed and D between C_f and C_r is reduced to 10 m then C_r should brake.

Based on S_3: If the distance (D) between the C_f and C_r is reduced to less than 30 m as the C_f decreases its speed then C_r should also decreases its speed. The contracts extracting from the R_1 are as follows:

C_1: First contract is created based on R_1 as the C_r increases its speed (this contract is revitalized in case when the scenario S_1 can occur).

Assumption: C_f should accelerate if the distance between C_f and C_r is less than 30 m.

Guarantee: A safe distance of 30 m should be maintained between C_f and C_r.

C_2: Second contract is created based on R_1 as C_f decreases its speed (this contract is revitalized in case when the scenario S_2 can occur).

Assumption: C_r will brake while the distance between C_f and C_r is less than 10 m

Guarantee: A safe distance of 10 m should be maintained between C_f and C_r.

C_3: Third contract is created based on R_1 as C_f decreases its speed (this contract is revitalized in case when the scenario S_3 can occur).

Assumption: C_r will reduce its speed when the minimum distance between C_r and C_f approaches to 30 m.

Guarantee: Safe distance of 30 m should be maintained between C_f and C_r.

Implementation of (Controller Implementing) Contracts in Simulation. The implementation of (controllers implementing) contracts in CARLA Scenic (along with the code snippets) is elaborated in Sect. 4. The controller implementing contract C_1 activates its collision avoidance mechanism when there is a probability of collision occurrence in the scenario S_1. Same procedure is applied for C_2 and C_3 for scenarios S_2, S_3, S_4, S_5, and S_6. However, the proposed implementation in this context not only "employs the contract" but also selects the particular implementations satisfying the contract i.e. changing the current speed of AV when there is a chance of collision. This distinction is fundamental to CBD in the context of our work.

Results. This section will demonstrate the results in detail. Each scenario runs multiple times and simulation results in CARLA before and after the implementation of (controller implementing) contracts are analyzed.

4 Running CARLA Simulation

As mentioned in Sect. 3 that Scenic is used with the CARLA simulator for implementing 1. the scenes, 2. the created scenarios and 3. the impact of the contract implementation wrt. collision avoidance. Next few sections will elaborate these steps in detail.

4.1 Creation of a Scene in Scenic

Scenic allows to create an environment E, by using different maps that are present in the repository of CARLA. These maps has urban areas along with different objects (which the user can add to) including buildings, trees, walls, street lights etc. Map supported CARLA libraries (a library containing maps) are available at [12].

4.2 Creation of Scenarios in Scenic

After creating environment E in Scenic as explained in Sect. 4.1. E allows the user to add the vehicles according to the requirement of defined scenarios. In our case we have created two AVs (C_f and C_r) for all proposed scenarios.

Scenario 1 (S_1): The scenario S_1 is elaborated in the Sect. 3. If the rear AV (C_r) increases its speed (V_f) then front AV (C_f) should accelerate. Therefore, in this scenario (S_1) the (controller implementing) contract C_1 is implemented to avoid the collision. The implementation of controller implementing C_1 for S_1 in Scenic is shown in Fig. 3.

Scenario 2 (S_2): Scenario S_2 is elaborated in the Sect. 3. If the front AV (C_f) decreases its speed (V_f) (due to the signal) then rear AV (C_r) should also reduce its speed (V_r). Therefore, in S_2 the C_2 (controllers implementing contract C_2) is implemented to avoid the collision (when the safe distance between C_f and C_r becomes less than 10 m or 30 m, C_r should minimize it speed first and then should brake due to the possibility of collision occurrence). The implementation of controller implementing C_2 in Scenic is shown in Fig. 4.

Scenario 3 (S_3): Scenario S_3 is elaborated in the Sect. 3. If front AV (C_f) decreases its speed V_f then rear AV (C_r) should also reduce its speed V_r. Therefore, in this scenario (S_3) the C_3 (controllers implementing contract C_3) is implemented to avoid the collision (when the safe distance D between C_f and C_r becomes less than 30 m). The implementation of C_3 is almost similar to the implementation shown in Fig. 4

```
#S1: After covering distance of 5km,
Cr increases its speed and D between
the Cf and Cr is less than 30m.
#S1 can be avoided by implementing C1
which allows the Cf to accelerate.
#C1
#Assumption: Cf should accelerate if
the distance between Cf and Cr is less
than 30m
#Guarantee: Safe distance of 30m
should maintained between Cf and  Cr
Distance = 30 #meters
Cf_Speed=30  #km/hr
Cf_Speed_New= Cf_Speed+40

behavior Cf(Cf_Speed):
if(distancefromselfto Cr)> Distance:
require
self.distanceToClosest(Object) <5
  do FollowLaneBehavior(Cf_Speed)
 else:
  do FollowLaneBehavior(Cf_Speed_New)
  do CollisionAvoidance()
  do GetBackOntoRoad()
```

Fig. 3. Code snippet C_1 for S_1

```
#S2: After covering distance of 5km,Cf
decreases its speed and D between Cf and
Cr is now less than 10m.
#S2 can be avoided by implementing C2
which allows the Cr to minimize its speed
(in case of 30m safe distance) and brakes
when the safe distance is 10m.
#C2
#Assumption: Cr will minimize its speed
and brake in case when the distance
between Cf and Cr is less than 10m
#Guarantee: Safe distance of 10m should
maintained between Cf and Cr

Distance = 30 #meters
Cr_Speed= 30  #km/hr
Cr_Speednew= 10  #km/hr

behavior Cr(Cr_Speed):

try:
  do
  FollowLaneBehavior(Cr_SPEED)
 interrupt
  when withinDistanceToAnyObjs(self,30):
   take SetSpeedAction(Cr_Speednew)
  when withinDistanceToAnyObjs(self,10):
   take SetBrakeAction(BRAKE_ACTION)
```

Fig. 4. Code snippet C_2 for S_2

Scenario 4 (S_4): Scenario S_4 is based on the scenario S_1 with different parameters for speed and safe distance. However, the A/G based contracts are not changed. Both AVs have the same speed while the safe distance is 10 m. These specifications of S_4 are implemented for contracts C_2 (used by S_5 based on S_2) and C_3 (used by S_6 based on S_3) so that the varied results are generated and analyzed.

5 Results

This section elaborates the results generated by running the CARLA simulation. Time taken by each scenario is represented by TS. Collision occurrence time (considering each scenario one by one) without the implementation of controllers implementing contracts is represented by CT. Collision occurrence time with the implementation of controllers implementing contracts is represented by CTC. Safe time with controllers implementing contracts is defined as the time until the AVs drive safely and collision do not occur, is represented by STC. Safe time without controllers implementing contracts is represented by ST. BC represents a mechanism (a built-in mechanism provided by CARLA) that avoids the collision of AVs with fences, walls and other objects on the road. $BC = 0$ means that there is no basic collision implemented in the simulation while $BC = 1$ means that basic collision detection mechanism is already employed in the simulation. The calculated time in all tables is represented in minutes.

Table 1. Scenarios without contracts

Scenarios	TS	CT	ST	BC
S_1 without C_1	–	–	–	–
Run 1 (R_{11})	10	3	2	0
Run 2 (R_{12})	10	4	3	0
Run 3 (R_{13})	10	6	5	1
Run 4 (R_{14})	10	5	4	1
S_2 without C_2	–	–	–	–
Run 1 (R_{15})	10	3	2	0
Run 2 (R_{16})	10	4	3	0
Run 3 (R_{17})	10	5	4	1
Run 4 (R_{18})	10	4	3	1
S_3 without C_3	–	–	–	–
Run 1 (R_{19})	10	4	3	0
Run 2 (R_{20})	10	5	4	0
Run 3 (R_{21})	10	6	5	1
Run 4 (R_{22})	10	5	4	1

Table 2. Scenarios with contracts

Scenarios	TS	CTC	STC	BC
S_1 with C_1	–	–	–	–
Run 1 (R_{31})	10	5	4	0
Run 2 (R_{32})	10	6	4	0
Run 3 (R_{33})	**10**	**10**	**10**	**1**
Run 4 (R_{34})	**10**	**10**	**9**	**1**
S_2 with C_2	–	–	–	–
Run 1 (R_{35})	10	6	4	0
Run 2 (R_{36})	10	5	4	0
Run 3 (R_{37})	**10**	**10**	**10**	**1**
Run 4 (R_{38})	**10**	**10**	**10**	**1**
S_3 with C_3	–	–	–	–
Run 1 (R_{39})	10	6	5	0
Run 2 (R_{40})	10	5	4	0
Run 3 (R_{41})	**10**	**10**	**10**	**1**
Run 4 (R_{42})	**10**	**10**	**10**	**1**

Table 3. Scenarios without contracts

Scenarios	TS	CT	ST	BC
S_4 without C_1	–	–	–	–
Run 1 (R_{43})	10	1	0	0
Run 2 (R_{44})	10	1	0	0
Run 3 (R_{45})	10	2	1	1
Run 4 (R_{46})	10	1	0	1
S_5 without C_2	–	–	–	–
Run 1 (R_{47})	10	1	1	0
Run 2 (R_{48})	10	2	1	0
Run 3 (R_{49})	10	3	2	1
Run 4 (R_{50})	10	3	2	1
S_6 without C_3	–	–	–	–
Run 1 (R_{51})	10	3	2	0
Run 2 (R_{52})	10	2	1	0
Run 3 (R_{53})	10	3	2	1
Run 4 (R_{54})	10	3	2	1

Table 4. Scenarios with contracts

Scenarios	TS	CTC	STC	BC
S_4 with C_1	–	–	–	–
Run 1 (R_{55})	10	4	3	0
Run 2 (R_{56})	10	6	5	0
Run 3(R_{57})	**10**	**10**	**10**	**1**
Run 4 (R_{58})	**10**	**10**	**10**	**1**
S_5 with C_2	–	–	–	–
Run 1 (R_{59})	10	5	4	0
Run 2 (R_{60})	10	4	3	0
Run 3 (R_{61})	**10**	**10**	**10**	**1**
Run 4 (R_{62})	**10**	**10**	**10**	**1**
S_6 with C_3	–	–	–	–
Run 1 (R_{63})	10	4	3	0
Run 2 (R_{64})	10	3	2	0
Run 3 (R_{65})	**10**	**10**	**10**	**1**
Run 4 (R_{66})	**10**	**10**	**10**	**1**

5.1 Case 1: Results Without Controller Implementing Contracts

In this case we are running all the scenarios without contracts while considering the two special cases for BC: 1. BC is implemented ($BC = 1$) and 2. BC is not implemented ($BC = 0$). Overall it is evident from the Table 1 that when different scenarios are run without contracts, collision will always occur specially in case when $BC = 0$. However, collisions arises some how later in the case of $BC = 1$ (visible from the ST increase) but still collision cannot be completely avoided. C_f and C_r do not collide with the fence or other objects on the road but they will collide with each other due to the fact that there is no implementation of the contracts. Therefore, when we consider the results in Table 3 where both AVs (C_f and C_r) are driving at the same speed. It is clearly evident that the collision occurs as soon as the simulation start running (in both cases when $BC = 1$ or $BC = 1$) and the overall safe time ST is dramatically decreased.

5.2 Case 2: Results with Controller Implementing Contracts

In this case we are running all the scenarios with controllers implementing the contracts while considering again the two possibilities for switching the basic collision avoidance for BC (contract-conformant control) on or off. Overall it is evident from the Table 2 that when the scenario is run with the controllers implementing the contracts, collision (then with the infrastructure) will still occur in case when $BC = 0$. However, no collisions occurs in the case $BC = 1$ as the controllers implement contracts for collision avoidance along with the basic collision mechanism, both working together to avoid collision (shown in the rows with bold). Therefore, when we consider the results in Table 4 where both AVs (C_f and C_r) are driving at the same speed. It is clearly evident that when the contracts are implemented along with the basic collision mechanism ($BC = 1$), collision can be completely avoided (as shown in the bold rows). But in all other cases the overall safe time without contracts (ST) and safe time with contracts (STC) is badly effected.

5.3 Discussion About Results

From Table 5 it is evident that out of the forty eight scenarios, only twelve scenarios (represented in blue color) featuring both 1. contract-conformant control and 2. basic collision avoidance mechanism allows the collision free driving of AVs. From these twelve scenarios it is evident that AV can avoid collision completely only in case when both features 1. contract-conformant control and 2. basic collision avoidance are implemented. Table 5 also considers the special scenarios in which the speed and safe distance of scenarios are changed dynamically to analyse the results. Although the overall results are similar in all scenarios, but the overall safe time without contracts (ST) and safe time with contracts (STC) is decreased in the case of special scenarios.

Table 5. Complete results

Scenarios	BC	Contract Applied	Collision Occurred
R_{11}			✓
R_{12}			✓
R_{13}	✓		✓
R_{14}	✓		✓
R_{15}			✓
R_{16}			✓
R_{17}	✓		✓
R_{18}	✓		✓
R_{19}			✓
R_{20}			✓
R_{21}	✓		✓
R_{22}	✓		✓
R_{31}		✓	✓
R_{32}		✓	✓
R_{33}	✓	✓	
R_{34}	✓	✓	
R_{35}		✓	✓
R_{36}		✓	✓
R_{37}	✓	✓	
R_{38}	✓	✓	
R_{39}		✓	✓
R_{40}		✓	✓
R_{41}	✓	✓	
R_{42}	✓	✓	
Special Scenarios			
R_{43}			✓
R_{44}			✓
R_{45}	✓		✓
R_{46}	✓		✓
R_{47}			✓
R_{48}			✓
R_{49}	✓		✓
R_{50}	✓		✓
R_{51}			✓
R_{52}			✓
R_{53}	✓		✓
R_{54}	✓		✓
R_{55}		✓	✓
R_{56}		✓	✓
R_{57}	✓	✓	
R_{58}	✓	✓	
R_{59}		✓	✓
R_{60}		✓	✓
R_{61}	✓	✓	
R_{62}	✓	✓	
R_{63}		✓	✓
R_{64}		✓	✓
R_{65}	✓	✓	
R_{66}	✓	✓	

6 Conclusion and Future Work

There is high speculation in the domain of AVs when the apprehensions related to safety and reliability are considered. However, the preeminent objective in this context is to enhance the safety of AVs by mitigating the risk of accidents (specifically when self-driving AVs are fully deployed on the roads in near future). The proposed technique in our work is based on capturing scenario-oriented A/G contracts and provides a methodology for systematically deriving collision avoidance maneuvers effective during driving. Our approach is based on 1. the specifications of relevant scenarios (taken as test cases), 2. elucidation of specific contracts for each scenario and implementation of a contract-conformant control for collision avoidance, 3. statistical testing of the contract-conformant control in work by simulating each scenario in CARLA Scenic. It is evident from the results that controllers derived from A/G-based contracts improve the overall safety of the AV by avoiding collision by providing safe driving maneuvers. For the future work, we will consider more diverse scenarios based on 1. dense traffic, 2. pedestrians, 3. human-operated vehicles, and 4. cyclists. This work will allow us to create more diverse scenario-oriented controllers implementing contracts and thereby reinforce the safety and reliability of AVs in cases where there is a high risk of an accident. Also, comparison to other approaches towards scenario-based safety assurance is subject to the part of our future research.

Acknowledgement. The research presented herein has been funded by the State of Lower Saxony as part of the collaborative research action Zukunftslabor Mobilität.

References

1. Kalra, N., Paddock, S.M.: Driving to safety: how many miles of driving would it take to demonstrate autonomous vehicle reliability? Transp. Res. Part A Policy Pract. **94**, 182–193 (2016)
2. Seshia, S.A., Hu, S., Li, W., Zhu, Q.: Design automation of cyber-physical systems: challenges, advances, and opportunities. IEEE Trans. Comput. Aided Des. Integr. Circuits Syst. **36**(9), 1421–1434 (2016)
3. Damm, W., Votintseva, A., Metzner, A., Josko, B., Peikenkamp, T., Böde, E.: Boosting re-use of embedded automotive applications through rich components. In: Proceedings of Foundations of Interface Technologies 2005 (2005)
4. Bak, S., Chaki, S.: Verifying cyber-physical systems by combining software model checking with hybrid systems reachability. In: Proceedings of the 13th International Conference on Embedded Software, pp. 1–10 (2016)
5. Benveniste, A., et al.: Contracts for system design. Ph.D. thesis, Inria (2012)
6. Fremont, D.J., et al.: Formal scenario-based testing of autonomous vehicles: from simulation to the real world. In: 2020 IEEE 23rd International Conference on Intelligent Transportation Systems (ITSC), pp. 1–8. IEEE (2020)
7. Majzik, I., et al.: Towards system-level testing with coverage guarantees for autonomous vehicles. In: 2019 ACM/IEEE 22nd International Conference on Model Driven Engineering Languages and Systems (MODELS), pp. 89–94. IEEE (2019)

8. Shalev-Shwartz, S., Shammah, S., Shashua, A.: On a formal model of safe and scalable self-driving cars. arXiv preprint arXiv:1708.06374 (2017)
9. Tuncali, C.E., Fainekos, G., Ito, H., Kapinski, J.: Sim-ATAV: simulation-based adversarial testing framework for autonomous vehicles. In: Proceedings of the 21st International Conference on Hybrid Systems: Computation and Control (part of CPS Week), pp. 283–284 (2018)
10. Fremont, D.J., et al.: Scenic: a language for scenario specification and scene generation. In: Proceedings of the 40th ACM SIGPLAN Conference on Programming Language Design and Implementation, pp. 63–78 (2019)
11. Dosovitskiy, A., Ros, G., Codevilla, F., Lopez, A., Koltun, V.: CARLA: an open urban driving simulator. In: Conference on Robot Learning, pp. 1–16. PMLR (2017)
12. Fremont, D.J., et al.: Scenic: a language for scenario specification and data generation. arXiv preprint arXiv:2010.06580 (2020)

Check for updates

Dynamic Imputation Methodology for Multi-source Streaming Mobility Data

Michiel Dhont[1,2](✉), Elena Tsiporkova[1], and Nicolás González-Deleito[1]

[1] EluciDATA Lab of Sirris, BluePoint Brussels, Bd A. Reyerslaan 80, 1030 Brussels, Belgium
{michiel.dhont,elena.tsiporkova,nicolas.gonzalez}@sirris.be
[2] Department of Electronics and Information Processing (ETRO), Vrije Universiteit Brussel, Pleinlaan 2, 1050 Brussels, Belgium

Abstract. The road network is becoming increasingly equipped with a multitude of sensors, monitoring a wide range of operating and contextual parameters. The availability of real-time sensor data enables the realisation of diverse data-driven applications, e.g., anomaly detection, identification of insightful patterns, monitoring the evolution of relevant trends in time and delivery of actionable decision support. However, such streaming data might contain vast amounts of missing values depending on the application. This makes it very challenging, if not impossible, to fully exploit the potential of data analysis and machine learning for these data sources, and in particular real-time analysis is not feasible. We propose in this paper an imputation methodology dedicated to multi-source streaming data, with a focus on the mobility domain. The proposed approach is based on spatio-temporal profiling of the streaming behaviour derived from historical data via non-negative matrix factorisation. The profiling method takes advantage of an adaptive segmentation strategy splitting the data into rolling time windows (chunks) allowing to use the limited non-missing data as optimally as possible. The identified profiles allow to devise a dynamic and scalable imputation strategy, which is able to reliably estimate incoming missing values in streaming data as soon as they arrive.

Keywords: Data imputation · Matrix factorisation · Streaming data · Vehicle counts

1 Introduction

There is an increasing trend of connecting devices (e.g., smart watches, smart household appliances and ANPR cameras) and industrial machinery (e.g., compressors, fleet tracking and melting furnaces) to the internet [20]. Since these

This research was subsidised through the project MISTic by the Brussels-Capital Region - Innoviris and received funding from the Flemish Government (AI Research Program).

Y. Bie et al. (Eds.): KES-STS 2022, SIST 304, pp. 184–198, 2022.
https://doi.org/10.1007/978-981-19-2813-0_19

assets are connected over a network, their data can be inspected in real-time. This real-time aspect opens a wide range of opportunities as it allows for early and continuous monitoring of trends and anomalies. By deriving an accurate view of the latest state of affairs at all time, real-time data-driven decision support applications can be developed.

Unfortunately, exploiting directly streaming data on the fly is not straightforward since it is often suffering from poor quality mostly due to incompleteness (e.g., in the mobility domain [16,25]). In practice, data capturing implementations are often rapid/experimental, i.e., sensors are quickly deployed to gather data. Moreover, these deployments can be continuously expanding in terms of additional sensors which are installed in order to capture supplementary information. Last, the devices may often be located in difficult to access or widely scattered locations (e.g., inductive loops under the asphalt of a busy road), making it difficult to do good maintenance (e.g., replace broken sensors). Due to these reasons, it is almost inevitable to have some flaws in the data capturing process because of technical failures. The latter may result in various types of missing data values: randomly scattered missing values, a single sensor missing over a longer period, and relatively short moments in time when no values are available at all.

Particularly in the case of streaming data, it is often important to be able to impute missing values as soon as a limited amount of values are available. However, due to the highly dynamic nature of the data and the frequent occurrences of long sequences of subsequent missing values, such an imputation is very challenging. For this reason, imputation approaches need to be able to learn from the past, capturing prototypical behaviour via adequate profiling mechanisms. Although there are many imputation algorithms proposed in the literature, some interesting opportunities have not yet been explored. Especially in the context of continuously incoming multi-source streaming (mobility) data, there is a lot of room for improvement.

Some imputation algorithms are developed with the specific goal to work on a continuous stream of data, while other algorithms try to capture the spatial and/or temporal patterns on a static data set. Although some of these imputation algorithms yield good results, a hybrid combination of both would be more relevant in practice since data is often both continuously incoming and exhibits multi-source correlations. In this paper, we propose a novel imputation methodology for multi-source data, capable of handling continuous streaming data and validated on real-life vehicle counting data. This approach is partially inspired by the spatio-temporal fingerprinting approach which was proposed in [6] and was originally used for the purpose of performance profiling. In this paper, we exploit cleverly its characterisation potential to dynamically detect latent spatial and temporal structures in multi-source data for the purpose of missing data imputation.

The remaining of this paper is organised as follows: Section 2 focuses on existing related approaches in the literature, while Sect. 3 outlines in detail the proposed novel imputation methodology. Section 4 covers the obtained validation results of a use case on mobility data. Finally, in Sect. 5, some concluding remarks are given.

2 Related Work

In this section, the state of the art in recent imputation strategies and approaches is discussed. First, in Sect. 2.1, existing imputation techniques for time series are briefly summarised. Next, in Sect. 2.2, the most relevant latest developments in streaming data imputation approaches are discussed.

2.1 Time Series Data Imputation

Existing imputation approaches for time series data can be divided into three different classes: interpolation, statistical learning and prediction [15]. The first class, interpolation, is the most straightforward imputation method, as it approximates the missing data by fitting a curve on top of the available data. The curve aims to define the sequence of data points by a linear or polynomial function, allowing to estimate unknown values [24]. Secondly, statistical learning-based imputation approaches aim to encapsulate statistical features of the data into a model. The latter could be achieved for example by applying the k-nearest neighbours approach [4,15,17], where estimations are made based on the k most similar situations, or a principal component analysis [18,22], where an expectation-maximisation algorithm is used to estimate values of missing data points. Finally, prediction-based imputation approaches aim to capture the temporal relationship within time series. These imputation algorithms are developed to find long term and short term relations, giving an idea about what value to expect next. The autoregressive integrated moving average method [13,28] and Bayesian networks [9,26] are two methods that belong to the category of prediction-based imputation approaches.

More recently, factorisation techniques are used to impute missing values in matrix-like data sets. Completion of matrix-like data has shown to be relevant in many applications, such as image inpainting [14] and recommender systems [21]. For example, singular value decomposition can be used for matrix completion [7]. Bao et al. [2] illustrate how this approach can be applied on a multi-variate time series data set, where each row represents a time series for a different sensor and each column captures exactly one value for each sensor at a particular moment in time. Apart from the good imputation results, the non-parametric approach allows for reliable data imputation without user intervention. Note that existing factorisation imputation techniques are only designed to fill in gaps within a matrix. In case of streaming data, new unseen columns (i.e., moments in time) are continuously added to the data matrix. Consequently, the full factorisation must be recalculated each time new columns with missing values are added to the back of the matrix. For high-frequency multi-variate data streams, such an approach would be highly inefficient and therefore unusable in real-time. Section 2.2 gives an overview on the latest developments on streaming data imputation.

2.2 Streaming Data Imputation

The increasing availability of real-time sensor data, opens a wide range of new opportunities to instantaneously analyse the data and provide data-driven decision-making. However, in the case of low-quality data, this means that data imputation strategies also need to be adapted to run efficiently on new chunks of incoming data. Coupled to this, data imputation algorithms should be able to define a level of certainty in order to express how reliable the imputation results are.

Depending on the use case, different streaming imputation strategies are appropriate. In situations where computational power is limited (think of data imputation on the edge) or a short latency is crucial, it is good practice to construct a fixed imputation model on beforehand which allows for efficient imputation on continuously incoming data. Following this approach, Fountas and Kolomvatsos propose an ensemble correlation approach to identify the pairwise similarity between a number of different sensors (i.e., multi-variate time series). Missing values are imputed based on the values of the top-k correlated sensors, weighted by their correlation [8]. In [6], a similar approach is proposed, imputing missing values in an incremental way. First, missing data is imputed based on the top-k moments in time with the most similar non-missing values. Next, the remaining gaps are filled by use of the most similar larger periods of time (e.g., days).

If the data is more complex (e.g., new unseen or diverging data patterns may arise over time) or there is initially only a limited amount of historical data available, a continuous learning imputation approach is expected to be more appropriate. To give an example, in [19] a single factorisation to identify temporal features for historical missing data imputation is exploited. Then, an incremental learning scheme based on an autoregressive model is proposed, allowing for response forecasting based on the temporal features. In the study of Halder et al. [10], some problems with imbalanced data during data stream imputation are considered. To overcome these problems, an adaptive imputation approach is proposed which includes an oversampling method per chunk of streaming data and a fuzzy decomposition method to determine the interrelationship among instances. Despite the good results on imbalanced data sets, this approach has some performance issues in the case of noisy data, which is rather crucial in a real-wold context. Furthermore, none of these methods are able to store and recognise historically occurred relations between sensors. For instance, imagine a multivariate time series that counts the number of vehicles on a number of streets close to a charging bridge over a canal. Whether this bridge has opened (and influenced the traffic flow) during a gap of missing data, is impossible to know (i.e., impute) based on only the time aspect of one sensor. In such a case, the imputation method should be able to dynamically recognise the situation based on the other sensors at that moment in time. The novel imputation method we propose in this work is able to deal with such situations and also tackles most of the other shortcomings of the related works discussed above. More specific, our approach can efficiently impute multi-variate streaming data, while still considering both temporal and spatial relations using a dynamic profiling methodology.

3 Materials and Methods

In this section the building blocks of our novel dynamic profiling and imputation methodology are discussed. Section 3.1 outlines the concept of matrix factorisation, while Sect. 3.2 is devoted to the description of the spatio-temporal profiling. Next, Sect. 3.3 explains how to exploit the profiles to impute missing values. Section 3.4 indicates how to deploy this approach on real-time streaming data. Finally, Sect. 3.5 provides a description of the test data set and the used computer code.

3.1 Matrix Factorisation

Matrix factorisation is a discipline of linear algebra allowing to decompose a matrix into a product of matrices. One popular example of this approach is the **singular value decomposition** (SVD). Consider a matrix $\boldsymbol{X} \in \mathbb{C}^{M \times N}$. By the use of SVD, $\boldsymbol{X}$ can be factorised into $\boldsymbol{U\Sigma V}^T$, with $\boldsymbol{U}$ a unitary matrix $\in \mathbb{C}^{M \times M}$, $\boldsymbol{\Sigma}$ a rectangular diagonal matrix $\in \mathbb{R}_+^{M \times N}$ and $\boldsymbol{V}$ a unitary matrix $\in \mathbb{C}^{N \times N}$. SVD has many applications, such as solving homogeneous linear equations (e.g., [1]), total least squares minimisation (e.g., [27]) and low-rank matrix approximations (e.g., [23]).

Another factorisation approach is **non-negative matrix factorisation** (NMF). As the name reveals, this approach is designed to work with matrices containing only positive values. Consider a matrix $\boldsymbol{X} \in \mathbb{R}_+^{M \times N}$. NMF is able to approximate $\boldsymbol{X}$ by a product of two factors $\boldsymbol{WS}$, where $\boldsymbol{W} \in \mathbb{R}_+^{M \times R}$, $\boldsymbol{S} \in \mathbb{R}_+^{R \times N}$ and $R < \min(M, N)$. The smaller the value of parameter R, the greater the dimensionality reduction performed, at the expense of the reconstruction error for $\boldsymbol{X}$. In contrast to SVD, NMF is an approximation for which no exact solution exists. There are multiple heuristic algorithms developed to find $\boldsymbol{W}$ and $\boldsymbol{S}$ [11]. In our approach, the fast hierarchical alternating least squares (Fast HALS) algorithm is used [5]. Compared to SVD, the added value of NMF is the constraint of having only positive values in matrices $\boldsymbol{W}$ and $\boldsymbol{S}$. Depending on the nature of the data from the original matrix, it often is a more natural process to decompose it into positive factors [12]. To reconstruct the original matrix $\boldsymbol{X}$, no element-wise subtractions need to be performed.

3.2 Spatio-Temporal Profiling

Spatio-temporal profiling is an essential prerequisite for our imputation strategy. It allows to extract latent spatial and temporal patterns from historical data, which are subsequently used by the imputation algorithm. Consider a matrix representing data from a multi-variate time series as visualised in Fig. 1(a). In such a matrix, each row represents one of the M different parameters (e.g., sensor values for different locations). Each column represents a consecutive timestamp (e.g., one value per minute). Considering we are working with streaming data, the time dimension is infinite.

Adaptive Dynamic Segmentation. To obtain resilient profiles, gaps of missing data are avoided during the profiling procedure. The first step in that procedure is to extract chunks of data with a fixed time window of N timestamps. The width N of each time window should be large enough so that meaningful temporal patterns can be identified, but small enough so that enough chunks of data without missing values can be found. The fulfilment of these requirements are dependent on the use case of interest. The data chunk extraction happens by chronologically looping over all timestamps $(t_1, t_2 \cdots, t_T)$, with t_1 the oldest timestamp and t_T the most recent timestamp. A chunk is only selected if it contains no missing values. Each time a chunk is selected, a number of timestamps is skipped before proceeding with the selection of the next chunk in order to avoid excessive overlap between chunks. As a rule of thumb, we do a forward jump of $\frac{1}{3}N$ timestamps in order to have a maximum overlap of $\frac{2}{3}N$ timestamps between chunks. In Fig. 1(b), the selected chunks from Fig. 1(a) are visualised. The overlap between the different chunks is essential for being able to capture transitions between different patterns in time.

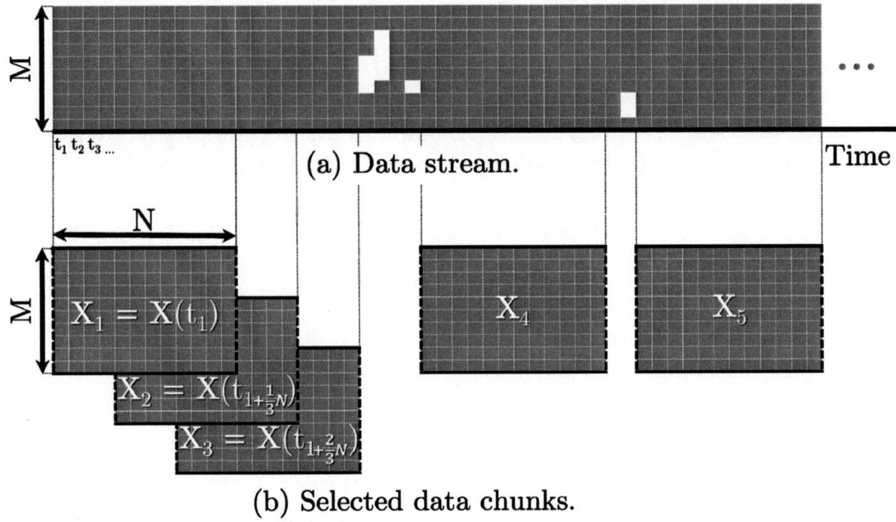

Fig. 1. Selection of data chucks for profiling.

In the second step, all selected data chunks are stacked on top of each other:

$$\boldsymbol{X} = \begin{bmatrix} \boldsymbol{X}_1 \\ \boldsymbol{X}_2 \\ \vdots \\ \boldsymbol{X}_L \end{bmatrix} \in \mathbb{R}_+^{(ML)\times N}, \tag{1}$$

where $\boldsymbol{X}_i$ represents the ith selected chunk and L the total number of selected chunks.

Temporal Profiles Extraction. Temporal profiles are extracted from the stacked matrix $\boldsymbol{X}$ by applying a decomposition method. In this paper, NMF is used due to the non-negative properties that are suitable for the use case data and to facilitate interpretation. In this way, matrices $\boldsymbol{W}$ and $\boldsymbol{S}$ are obtained as shown in Eq. (2). Conceptually, each row of matrix $\boldsymbol{S}$ can be interpreted as a temporal profile while matrix $\boldsymbol{W}$ represents the weights, which can be used to reconstruct $\boldsymbol{X}$ thanks to the temporal building blocks from $\boldsymbol{S}$.

$$\boldsymbol{X} \approx \boldsymbol{W}\boldsymbol{S}, \tag{2}$$

$$\boldsymbol{X} \approx \begin{bmatrix} \boldsymbol{W}_1 \\ \boldsymbol{W}_2 \\ \vdots \\ \boldsymbol{W}_L \end{bmatrix} \boldsymbol{S}. \tag{3}$$

These matrices are as follows: $\boldsymbol{W} \in \mathbb{R}_+^{(ML)\times R}$ and can be evenly split into L sub-matrices $\boldsymbol{W}_i \in \mathbb{R}_+^{M\times R}$, with $i = 1, 2, \ldots, L$ (see Eq. (3)), $\boldsymbol{S} \in \mathbb{R}_+^{R\times N}$, and $R \in \mathbb{N}_+$ being a hyperparameter representing the amount of temporal profiles, as explained in Sect. 3.1. Moreover, each chunk $\boldsymbol{X}_i$ can be approximated by a weighted sum of the temporal profiles as shown in Equation (4).

$$\boldsymbol{X}_i \approx \boldsymbol{W}_i\boldsymbol{S}, \quad \text{with } 1 \le i \le L. \tag{4}$$

Alternatively, SVD or any other decomposition technique could be used depending on the properties of the use case (e.g., if values can be negative). Note that in the case of SVD three matrices are obtained (see Sect. 3.1). However, in this situation, matrices $\boldsymbol{U}$ and $\boldsymbol{\Sigma}$ should be multiplied to replace the weight matrix $\boldsymbol{W}$, while $\boldsymbol{V}^T$ can directly be used as the temporal profile matrix $\boldsymbol{S}$.

Spatial Profiles Extraction. The decomposition of matrix $\boldsymbol{X}$ above via NMF resulted in latent temporal profiles and corresponding weights. The weight matrix $\boldsymbol{W}$ is of interest for further decomposition since it contains useful relational information between the different data sources. In situations where each parameter represents a sensor at a different location, this can be interpreted as the spatial relationship. To extract those relations, each individual weight matrix $\boldsymbol{W}_i$ is first transposed. Then, a modified weight matrix $\boldsymbol{W}'$ is constructed that vertically stacks all individual transposed matrices:

$$\boldsymbol{W}' = \begin{bmatrix} \boldsymbol{W}_1^T \\ \boldsymbol{W}_2^T \\ \vdots \\ \boldsymbol{W}_L^T \end{bmatrix} \in \mathbb{R}_+^{(RL)\times M}. \tag{5}$$

Next, we approximate this modified weight matrix $\boldsymbol{W}'$ (by using a suitable factorisation approach) as the product of two matrices $\boldsymbol{V}$ and $\boldsymbol{U}$, as shown in Eq. (6). Assuming we again use NMF, both matrices will be non-negative. Similarly as above, the rows of the resulting matrix $\boldsymbol{U}$ can be interpreted as a set of prototypical spatial profiles, which can be used to reconstruct $\boldsymbol{W}'$ by the weights of matrix $\boldsymbol{V}$.

$$\boldsymbol{W}' \approx \boldsymbol{V}\boldsymbol{U}, \tag{6}$$

$$\boldsymbol{W}' \approx \begin{bmatrix} \boldsymbol{V}_1 \\ \boldsymbol{V}_2 \\ \vdots \\ \boldsymbol{V}_L \end{bmatrix} \boldsymbol{U}, \tag{7}$$

with $\boldsymbol{V} \in \mathbb{R}_+^{(RL)\times Q}$, $\boldsymbol{V}$ can be evenly split into L sub-matrices $\boldsymbol{V}_i \in \mathbb{R}_+^{R\times Q}$, with $i = 1, 2, \ldots, L$, $\boldsymbol{U} \in \mathbb{R}_+^{Q\times M}$, and $Q \in \mathbb{N}_+$ being a hyperparameter representing the amount of spatial profiles. Similarly, Eq. (7) can be split as follows:

$$\boldsymbol{W}_i^T \approx \boldsymbol{V}_i\boldsymbol{U}, \quad \text{with } 1 \leq i \leq L. \tag{8}$$

3.3 Estimation of Missing Values

Spatial and temporal profiles, as extracted in Sect. 3.2, contain very valuable information which can be used to estimate missing values. To do that, we combine Eq. (4) and (8) as follows:

$$\boldsymbol{X}_i \approx (\boldsymbol{V}_i\boldsymbol{U})^T\boldsymbol{S}. \tag{9}$$

In this equation, any chunk $\boldsymbol{X}_i$ is expressed as the weighted combination of the temporal and spatial profiles using the weight matrix $\boldsymbol{V}_i$. Since the temporal and spatial profiles contain the latent building blocks for any period in time, this equation is assumed to also hold for any chunk with missing data ($\boldsymbol{X}_{missing}$). In that case, some values of $\boldsymbol{X}_{missing}$, as well as $\boldsymbol{V}_{missing}$, are unknown. Consequently, these unknown values can be heuristically found as a continuous optimisation problem. Technically, this can be achieved by minimising the squared error between values from the resulting matrices from the left and right sides of Eq. (9), as shown in Eq. (10):

$$\begin{aligned}
\min_{\chi, \boldsymbol{V}_{missing}} \ : \ & \left(\boldsymbol{X}_{missing[1,1]} - \widehat{\boldsymbol{X}}_{[1,1]}\right)^2 + \cdots + (\boldsymbol{X}_{missing[1,N]} - \widehat{\boldsymbol{X}}_{[1,N]})^2 \\
& + \left(\boldsymbol{X}_{missing[2,1]} - \widehat{\boldsymbol{X}}_{[2,1]}\right)^2 + \cdots + (\boldsymbol{X}_{missing[2,N]} - \widehat{\boldsymbol{X}}_{[2,N]})^2 \\
& \quad \vdots \\
& + \left(\boldsymbol{X}_{missing[M,1]} - \widehat{\boldsymbol{X}}_{[M,1]}\right)^2 + \cdots + (\boldsymbol{X}_{missing[M,N]} - \widehat{\boldsymbol{X}}_{[M,N]})^2 \\
\text{s.t.} : \ & \widehat{\boldsymbol{X}} = (\boldsymbol{V}_{missing}\boldsymbol{U})^T \boldsymbol{S} \\
& \chi := \{ \boldsymbol{X}_{missing[i,j]} \mid [i,j] \ is \ missing \ in \ \boldsymbol{X}_{missing} \} \\
& x \geq 0 \qquad\qquad for \ x \in \chi \\
& \boldsymbol{V}_{missing[i,j]} \geq 0 \quad for \ 1 \leq i \leq Q \ and \ 1 \leq j \leq R,
\end{aligned} \tag{10}$$

with indices between squared brackets representing the coordinates of one value within a matrix, e.g., $\boldsymbol{X}_{missing[i,j]}$ being the value in matrix $\boldsymbol{X}_{missing}$ at row i and column j.

3.4 Imputation Strategy

The design of our novel imputation approach allows for data imputation on both historical and streaming data. The approach is focused on mobility data due to its strong spatial and temporal dependencies. However, it can be used in other domains that exhibit such strong dependencies. The imputation workflow's steps are as follows:

1. **Composition of training data repository.** Starting from historical data covering a sufficiently long time period allowing to capture all possible temporal and spatial patterns, a representative training data repository of only complete data chunks is composed following the segmentation approach in Sect. 3.2.
2. **Extraction of spatio-temporal profiles.** Following the two-step process described in Sect. 3.2, prototypical spatio-temporal profiles are extracted from matrix $\boldsymbol{X}$, constructed by stacking vertically the data chunks from the training repository.
3. **Imputation.** To impute data, chunks with missing data are extracted in the same way as in Fig. 1. However, a candidate data chunk $\boldsymbol{X}(t_i)$ is now only selected if it contains at least one missing value. Subsequently, the missing values in each data chunk are estimated as outlined in Sect. 3.3. Since chunks are allowed to overlap with $\frac{2}{3}N$ timestamps, each missing data point occurs in exactly 3 chunks. The relative position of a particular missing value in a chunk has an influence on the matched temporal profiles since each time window captures a different part in time, giving thus slightly deviating estimations. To obtain the most resilient imputation, the average of all three estimations is used to finally impute the missing value.

- For *historical data*, all missing values are imputed at once following the process described above.
- For real-time *streaming data*, at any moment a chunk with missing data is detected and selected, estimations are immediately computed for missing values in that chunk. Because the average of the estimation for the 3 overlapping incomplete chunks is used as final imputation value, the real-time imputation faces a latency of up to one time window (N timestamps). In parallel, one should monitor for concept drift since streaming data might be capturing a deviating or changing phenomenon. In that case, the spatial and temporal profiles need to be updated. To ease this update, chunks with no missing values should be identified as they are encountered and stored in the training data repository in order to be used later on.

3.5 Data and Computer Code

To illustrate the novel imputation approach proposed in this work, experiments have been carried out on a real-world data set from the mobility domain. More specifically, vehicle counts from 16 automatic number-plate recognition (ANPR) cameras were used. As shown in Fig. 2, the cameras are situated in 8 different locations on a circumferential urban highway (i.e., the small ring of Brussels, Belgium), while each camera monitors traffic in one direction. The data covers a period of 20 months, from February 2020 until the end of 2021. Within this period, the amount of vehicles that passed by per minute is provided for each location. This data has been collected using the real-time open API of Brussels Mobility[1], the public administration responsible for the mobility infrastructure in Brussels.

An interesting aspect of this specific data set is that the 16 ANPR cameras are situated along one single road, half of them in each direction (see Fig. 2). Therefore, many vehicles traverse several, if not all, of the 8 locations in one direction, creating a flow of vehicles. Note however that, as only aggregated information about vehicle counts is available, it is not possible to track the trajectory followed by an individual vehicle. It is important to understand that the quality of this real-world data set is not very high. Over 23% of all values are missing, making further advanced analysis of this data not really feasible, unless an appropriate data imputation method could increase the completeness.

The implementation of the proposed methods was done in Python. The Python code can be provided on request.

4 Results and Benchmarking

To validate the proposed imputation method, vehicle counting data for 16 locations, as described in Sect. 3.5, is used. Section 4.1 explains the construction of the training and validation data sets. Next, in Sect. 4.1, the imputation results on the validation data sets are given. Finally, the imputation results are benchmarked in Sect. 4.2.

[1] https://data.mobility.brussels/traffic/api/counts/.

Fig. 2. Map of the ANPR cameras in Brussels, Belgium.

4.1 Validation Strategy

The original real-world data set as described in Sect. 3.5 is first smoothed such that a continuous flow of data points is obtained. The latter is achieved by assigning per minute the mean value of the (known) values for the last 15 min. In this way, time gaps of up to 14 min are filled while the data set's original granularity is preserved.

Validation Data Sets. To allow for a qualitative and objective validation of our imputation approach, the ground truth of the missing values needs to be known. Since the missing data rate of our original data set (>23%) is too high to introduce additional missing values, several validation data sets are constructed as follows. After smoothing, only the 34 days with no missing values are retained. Next, 5 validation data sets with missing values are composed by randomly removing data values from these 34 days. Each data set has a different level of missing values: 5%, 10%, 15%, 20% and 25%.

Training Data Sets. To derive representative spatio-temporal profiles, we need more than 34 full days. For this reason, a dedicated training data set is composed for each of the 5 validation data sets, containing the smoothed data of all 20 available months. The 34 full days are contained with their different amounts of added missing values across the different data sets.

To validate our imputation approach on *historical data*, the training data set allows for the derivation of spatio-temporal models from the full time period, while the validation data set is used to validate the imputation accuracy based on the ground truth of the 34 full days. For the validation of our approach on *streaming data*, we split both data sets in two parts. The first 9 months of the training set are used to derive the spatio-temporal models. Next, the (16) days in the validation set that fall after those 9 months are used to validate the imputation approach. In this way, we can test how well our historical spatio-temporal patterns can be used to impute future data.

Profile Extraction. As explained in Sect. 3.2, first the spatial and temporal profiles are extracted from the training set. The chunks $\boldsymbol{X}_i$ have a spatial dimension (M) of 16 and a temporal dimension of 3 h ($N = 180$). The latter was chosen as a trade-off between a higher chance to segment a complete time window (no missing values), while still capturing sequences that represent a meaningful tendency. In addition, the lower the value of N, the larger the training set of chunks becomes. Since we use NMF as decomposition approach, the rank hyperparameter has to be chosen for both the extraction of the spatial and the temporal profiles. As validation method for the rank of the temporal (R) and spatial (Q) profile extraction we used the explained variance, i.e., the ratio between the variance after reconstruction and the original variance. We considered a rank resulting in an explained variance of over 99% as fulfilling, although we experimented with some other levels of explained variance. To obtain an explained variance of 99%, an R value of 64 and a Q value of 14 were chosen. To estimate the imputation values, the limited-memory version with bound constrains of the Broyden-Fletcher-Goldfarb-Shanno algorithm (L-BFGS-B) was used [3]. As initial solution for $\boldsymbol{V}_{missing}^T$, a linear interpolation per vehicle counting location was used. As shown in Fig. 3, the amount of latent temporal and spatial profiles has a big impact on the root mean squared error (RMSE) for all of the 5 training and validation data sets. Note that the RMSE values are only based on the artificially removed data values in 34 out of the 708 days since we only know the ground truth of these values. Remarkably, the imputation approach gives better results in the cases where more data is missing. The latter might be attributed to the increased degrees of freedom, avoiding the optimisation algorithm to overfit on the non-missing values. This statement will of course not hold for more extreme ratios of missing values.

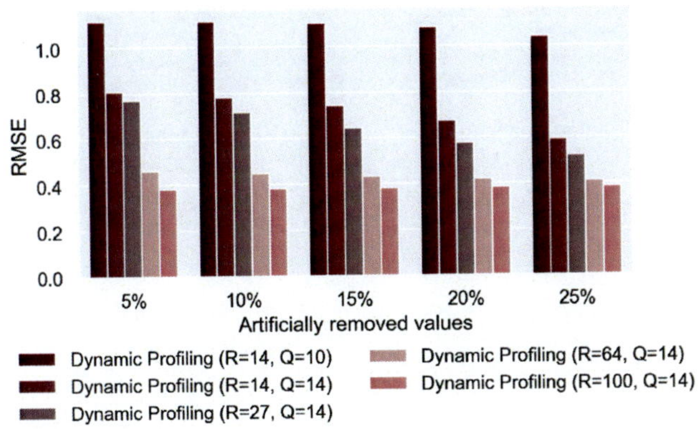

Fig. 3. RMSE values for various hyperparameters on different degrees of missing values.

4.2 Imputation Results

Figure 4a compares the RMSE of the two imputation approaches, in combination with the two best parameter settings from Fig. 3 for the validation strategy on historical data. As baseline imputation algorithm, we simply impute a missing value by copying the value from one week back at the same time. If that value would also be missing, we trace back in past weeks until a value is found. Traffic has a very clear weekly pattern and therefore this imputation method produces quite accurate imputation results as it can be witnessed in Fig. 4a (see "Preceding Weekday Imputation"). To compare the results with a state of the art imputation algorithm, the incremental spatio-temporal imputation method from [6] was chosen. Since the same data set was used in that paper, we reused the same hyperparameters. Compared to the baseline algorithm, only a small improvement was obtained for missing value rates of 5% and 10% as depicted in Fig. 4a (see "Incremental Imputation Method"). Figure 4a also illustrates that the two different versions of our dynamic profiling approach strongly outperform both alternative imputation approaches. Moreover, the strength of the dynamic profiling approach for higher missing rates is also very clearly demonstrated.

In Fig. 4b, the RMSE values for the validation strategy on streaming data are shown. Although this is a more difficult task, as confirmed by the increased RMSE values of the "Preceding Weekday Imputation" and "Incremental Imputation Method", the performance of the dynamic imputation methodology is equally as good as during the benchmarking on historical data imputation. This illustrates that our novel dynamic imputation method is able to robustly extract the latent spatio-temporal structures even from a reduced historical data set.

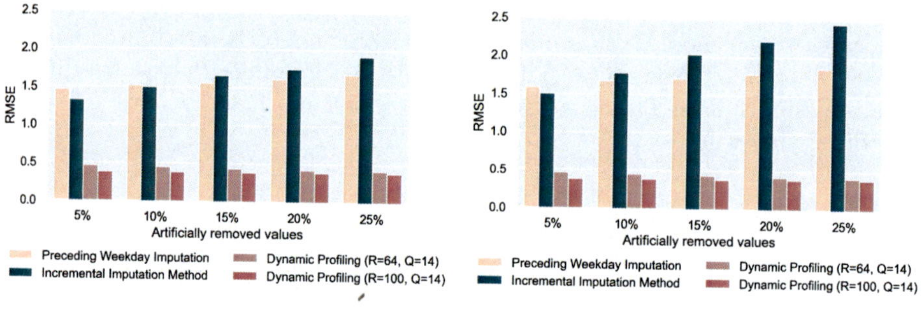

(a) Historical data benchmarking. (b) Streaming data benchmarking.

Fig. 4. Comparative results for various imputation strategies on different degrees of missing values.

5 Conclusion

In this work, we introduced a novel data imputation approach for multi-source streaming data. The method relies on spatio-temporal patterns, which are extracted via a double factorisation approach, and are able to encapsulate latent information structures in historical data. Real-world vehicle counting data has been used for the validation phase. The obtained results show that the approach performs extremely well for data sets with high rates of missing values (20–25%). The latter are very often detected in mobility data.

As future research we plan a further validation of the imputation approach by considering more advanced and realistic patterns for missing values, including a single sensor missing over a longer period and relatively short moments in time where no values are available at all. Our expectations are that the usage of spatio-temporal profiles might be even superior to alternative imputation methods as our approach can exploit patterns from both spatial and temporal dimensions simultaneously. Finally, we will try to improve the dynamic imputation methodology by experimenting with more intelligent initialisation strategies, as these can help the L-BFGS-B algorithm to converge faster and find better estimations.

References

1. Akritas, A., Malaschonok, G., Vigklas, P.: The SVD-fundamental theorem of linear algebra. Nonlinear Anal. Model. Control **11**(2), 123–136 (2006)
2. Bao, Z., Chang, G., Zhang, L., Chen, G., Zhang, S.: Filling missing values of multi-station GNSS coordinate time series based on matrix completion. Measurement **183**, 109862 (2021)
3. Byrd, R.H., Lu, P., Nocedal, J., Zhu, C.: A limited memory algorithm for bound constrained optimization. SIAM J. Sci. Comput. **16**(5), 1190–1208 (1995)
4. Chen, J., Shao, J.: Nearest neighbor imputation for survey data. J. Official Stat. **16**(2), 113 (2000)
5. Cichocki, A., Phan, A.H.: Fast local algorithms for large scale nonnegative matrix and tensor factorizations. IEICE Trans. Fundam. Electron. Commun. Comput. Sci. **92**(3), 708–721 (2009)
6. Dhont, M., Tsiporkova, E., González-Deleito, N.: Deriving spatio-temporal trajectory fingerprints from mobility data using non-negative matrix factorisation. In: 2021 International Conference on Data Mining Workshops (ICDMW), pp. 750–759. IEEE (2021)
7. Feuerverger, A., He, Y., Khatri, S.: Statistical significance of the Netflix challenge. Stat. Sci. **27**(2), 202–231 (2012)
8. Fountas, P., Kolomvatsos, K.: A continuous data imputation mechanism based on streams correlation. In: 2020 IEEE Symposium on Computers and Communications (ISCC), pp. 1–6. IEEE (2020)
9. Ghosh, B., Basu, B., O'Mahony, M.: Bayesian time-series model for short-term traffic flow forecasting. J. Transp. Eng. **133**(3), 180–189 (2007)
10. Halder, B., Ahmed, M.M., Amagasa, T., Isa, N.A.M., Faisal, R.H., Rahman, M., et al.: Missing information in imbalanced data stream: fuzzy adaptive imputation approach. Appl. Intell., 1–23 (2021)
11. Langville, A.N., Meyer, C.D., Albright, R., Cox, J., Duling, D.: Algorithms, initializations, and convergence for the nonnegative matrix factorization. arXiv preprint arXiv:1407.7299 (2014)
12. Lee, D.D., Seung, H.S.: Learning the parts of objects by non-negative matrix factorization. Nature **401**(6755), 788–791 (1999)

13. Lee, S., Fambro, D.B.: Application of subset autoregressive integrated moving average model for short-term freeway traffic volume forecasting. Transp. Res. Rec. **1678**(1), 179–188 (1999). https://doi.org/10.3141/1678-22
14. Li, X.P., Liu, Q., So, H.C.: Rank-one matrix approximation with l p-norm for image inpainting. IEEE Signal Process. Lett. **27**, 680–684 (2020)
15. Li, Y., Li, Z., Li, L.: Missing traffic data: comparison of imputation methods. IET Intel. Transport Syst. **8**(1), 51–57 (2014)
16. Nikfalazar, S., Yeh, C.-H., Bedingfield, S., Khorshidi, H.A.: A hybrid missing data imputation method for constructing city mobility indices. In: Islam, R., et al. (eds.) AusDM 2018. CCIS, vol. 996, pp. 135–148. Springer, Singapore (2019). https://doi.org/10.1007/978-981-13-6661-1_11
17. Oehmcke, S., Zielinski, O., Kramer, O.: KNN ensembles with penalized DTW for multivariate time series imputation. In: 2016 International Joint Conference on Neural Networks (IJCNN), pp. 2774–2781. IEEE (2016)
18. Qu, L., Li, L., Zhang, Y., Hu, J.: PPCA-based missing data imputation for traffic flow volume: a systematical approach. IEEE Trans. Intell. Transp. Syst. **10**(3), 512–522 (2009)
19. Ren, P., Chen, X., Sun, L., Sun, H.: Incremental Bayesian matrix/tensor learning for structural monitoring data imputation and response forecasting. Mech. Syst. Signal Process. **158** (2021)
20. Shi, Z., Chen, J., He, S.: DIY smart house: exploration and practice of IoT MOOC education. In: 2020 15th International Conference on Computer Science & Education (ICCSE), pp. 557–560. IEEE (2020)
21. Sun, S., et al.: Joint matrix factorization: a novel approach for recommender system. IEEE Access **8**, 224596–224607 (2020)
22. Tipping, M.E., Bishop, C.M.: Mixtures of probabilistic principal component analyzers. Neural Comput. **11**(2), 443–482 (1999)
23. Ye, J.: Generalized low rank approximations of matrices. Mach. Learn. **61**(1–3), 167–191 (2005)
24. Yin, W., Murray-Tuite, P., Rakha, H.: Imputing erroneous data of single-station loop detectors for nonincident conditions: comparison between temporal and spatial methods. J. Intell. Transp. Syst. **16**(3), 159–176 (2012)
25. Zafar, A., Kamran, M., Shad, S.A., Nisar, W.: A robust missing data-recovering technique for mobility data mining. Appl. Artif. Intell. **31**(5–6), 425–438 (2017)
26. Zhang, C., Sun, S., Yu, G.: A Bayesian network approach to time series forecasting of short-term traffic flows. In: Proceedings, The 7th International IEEE Conference on Intelligent Transportation Systems (IEEE Cat. No. 04TH8749), pp. 216–221. IEEE (2004)
27. Zhang, C., Chen, Q., Wang, M., Wei, S.: Optimised two-dimensional orthogonal matching pursuit algorithm via singular value decomposition. IET Signal Proc. **14**(10), 717–724 (2021)
28. Zhong, M., Sharma, S., Lingras, P.: Genetically designed models for accurate imputation of missing traffic counts. Transp. Res. Rec. **1879**(1), 71–79 (2004)

Author Index

Y. Bie et al. (Eds.): KES-STS 2022, SIST 304, pp. 199–200, 2022.
https://doi.org/10.1007/978-981-19-2813-0